COMPUTATIONAL APPROACHES IN CHEMINFORMATICS AND BIOINFORMATICS

COMPUTATIONAL APPROACHES IN CHEMINFORMATICS AND BIOINFORMATICS

Edited by

RAJARSHI GUHA
ANDREAS BENDER

WILEY

A JOHN WILEY & SONS, INC., PUBLICATION

Published by John Wiley & Sons, Inc., Hoboken, New Jersey.
Published simultaneously in Canada.

Library of Congress Cataloging-in-Publication Data:

Computational approaches in cheminformatics and bioinformatics / edited by Rajarshi Guha, Andreas Bender. – 1st ed.
p. cm.
Includes index.
ISBN 978-0-470-38441-1 (hardback)
1. Cheminformatics. 2. Bioinformatics. 3. Drugs–Research–Data processing. I. Guha, Rajarshi. II. Bender, Andreas.
QD39.3.E46C626 2012
615.10285–dc23

2011024792

Printed in the United States of America

10 9 8 7 6 5 4 3 2 1

CONTENTS

CONTRIBUTORS

ANDREAS BENDER, Unilever Centre for Molecular Science Informatics, Department of Chemistry, University of Cambridge, Cambridge, UK

KRISTIN P. BENNETT, Rensselaer Exploratory Center for Cheminformatics Research, Rensselaer Polytechnic Institute, Troy, New York

MICHAEL R. BERTHOLD, Universität Konstanz, Konstanz, Germany

EVAN E. BOLTON, National Center for Biotechnology Information, National Library of Medicine, National Institutes of Health, Bethesda, Maryland

CURT M. BRENEMAN, Rensselaer Exploratory Center for Cheminformatics Research, Rensselaer Polytechnic Institute, Troy, New York

STEPHEN H. BRYANT, National Center for Biotechnology Information, National Library of Medicine, National Institutes of Health, Bethesda, Maryland

VIVIEN CHAN, Oncology and Global Discovery Chemistry, Novartis Institutes for BioMedical Research, Emeryville, California

BEN CORNETT, Center for Proteomic Chemistry, Novartis Institutes for BioMedical Research, Cambridge, Massachusetts

SOURAV DAS, Rensselaer Exploratory Center for Cheminformatics Research, Rensselaer Polytechnic Institute, Troy, New York

JOHN W. DAVIES, Center for Proteomic Chemistry, Novartis Institutes for BioMedical Research, Cambridge, Massachusetts

MARTIN EKLUND, Department of Pharmaceutical Biosciences, Uppsala University, Uppsala, Sweden

JEAN-LOUP FAULON, Institute of Systems & Synthetic Biology, CNRS, University of Evry, France

ANGELO D. FAVIA, Drug Discovery and Development, Istituto Italiano di Tecnologia, Genoa, Italy

SHEKHAR GARDE, Rensselaer Exploratory Center for Cheminformatics Research, Rensselaer Polytechnic Institute, Troy, New York

RAHUL GODAWAT, Rensselaer Exploratory Center for Cheminformatics Research, Rensselaer Polytechnic Institute, Troy, New York

RAJARSHI GUHA, NIH Chemical Genomics Center, Rockville, Maryland

WOLF-D. IHLENFELDT, Xemistry GmbH, Lahntal, Germany

JEREMY L. JENKINS, Center for Proteomic Chemistry, Novartis Institutes for BioMedical Research, Cambridge, Massachusetts

JASON KONDRACKI, Global Discovery Chemistry, Novartis Institutes for BioMedical Research, Emeryville, California

MICHAEL KREIN, Rensselaer Exploratory Center for Cheminformatics Research, Rensselaer Polytechnic Institute, Troy, New York

MARIS LAPINS, Department of Pharmaceutical Biosciences, Uppsala University, Uppsala, Sweden

SHAWN MARTIN, Computer Science and Informatics, Sandia National Laboratories, Albuquerque, New Mexico

THORSTEN MEINL, Universität Konstanz, Konstanz, Germany

DMITRI MIKHAILOV, Center for Proteomic Chemistry, Novartis Institutes for BioMedical Research, Cambridge, Massachusetts

MILIND MISRA, Advanced Device Technologies, Sandia National Laboratories, Albuquerque, New Mexico

FLORIAN NIGSCH, Chemical Biology Informatics, Quantitative Biology, Departmental and Molecular Pathways, Novartis Institutes for BioMedical Research, Cambridge, Massachusetts

Irene Nobeli, Institute of Structural and Molecular Biology, Department of Biological Sciences, Birkbeck, University of London, London, UK

David E. Patterson, Vistamont Consultancy, Berkeley, California

Bernhard Rohde, Center for Proteomic Chemistry, Novartis Institutes for BioMedical Research, Novartis Pharma AG, Basel, Switzerland

Josef Scheiber, Center for Proteomic Research, Novartis Institutes for BioMedical Research, Cambridge, Massachusetts; currently at Pharma Research and Early Development Informatics, Pharma Research and Early Development, Roche Diagnostics GmbH, Penzberg, Germany

Ansgar Schuffenhauer, Center for Proteomic Chemistry, Novartis Institutes for BioMedical Research, Novartis Pharma AG, Basel, Switzerland

Ola Spjuth, Department of Pharmaceutical Biosciences, Uppsala University, Uppsala, Sweden

Sukumar N., Rensselaer Exploratory Center for Cheminformatics Research, Rensselaer Polytechnic Institute, Troy, New York

Paul A. Thiessen, National Center for Biotechnology Information, National Library of Medicine, National Institutes of Health, Bethesda, Maryland

Inna vitol, Rensselaer Exploratory Center for Cheminformatics Research, Rensselaer Polytechnic Institute, Troy, New York

Jarl E. S. Wikberg, Department of Pharmaceutical Biosciences, Uppsala University, Uppsala, Sweden

Bernd Wiswedel, Universität Konstanz, Konstanz, Germany

FOREWORD

The field of what we now refer to as chemoinformatics started some fifty years ago with the first attempts to search for substructural patterns in molecules[1] and to correlate biological activity with structural information.[2] Since then, chemoinformatics has evolved an entire range of tools and techniques for the discovery of novel molecules with important, and commercially valuable, properties.[3] To quote Paris's wide-ranging definition (as reported by Warr[4]), chemoinformatics encompasses "the design, creation, organization, management, retrieval, analysis, dissemination, visualization and use of chemical information" and is now one of the key tools used in the pharmaceutical industry for the discovery of new drugs. This has long been the most important application area, but chemoinformatics is also used extensively for the discovery of new insecticides, herbicides, and pesticides in the agrochemicals industry; it is also finding increasing application in market sectors such as foods and flavorings, personal products, and nutraceuticals.

All of these applications seek to make predictions about the biological activities of molecules from a knowledge of (primarily) their chemical structures, and the computational modeling involved hence requires a tight coupling of chemical and biological information. As Paolini et al. note: "correct compilation and integration of chemical and biological data is the foundation that transforms drug discovery into a knowledge based predictive science."[5] This book seeks to foster this integration by looking at some of the new ways in which researchers are seeking to exploit the implicit linkages that exist between chemistry and biology. The book's nine chapters cover four main areas: the data sources that are available for modeling and prediction purposes; developments of conventional quantitative structure–activity relationships (QSARs); computational tools for manipulating chemical and biological data; and novel ways of probing the interactions between small molecules and proteins. These four areas are reviewed below.

Thiessen et al. in Chapter 1 discuss PubChem, which is the largest Web-available database containing both chemical structures and associated bioactivity data, making it a key resource for computer-aided drug design. PubChem is unique in being an open repository for externally generated data; this is a very valuable characteristic but has required the development of sophisticated techniques for data normalization and searching to overcome the variations and errors that inevitably occur with data that have been donated from a wide range of sources. Bringing data together from different sources is also the focus of Chapter 2 by Jenkins et al. As the authors note, pharmaceutical companies are "drowning in data and thirsty for knowledge" and it is challenging task, even for the largest and best-resourced pharmaceutical companies, to integrate and exploit the huge volumes of disparate types of data necessary for successful drug discovery. These difficulties have been tackled in the Novartis Data Federation Initiative (NDFI), which has sought to provide a common data source for a huge range of complex scientific queries. The operation of NDFI is illustrated by proof-of-concept experiments using kinases, for which large amounts of both chemical and biological data are available.

The traditional computational approach to linking chemistry and biology is by the study of QSARs, where one probes the interaction of a set of (often closely related) molecules with a single biological target. In Chapter 3, Wikberg et al. describe proteochemometrics, which broadens the QSAR approach to allow it to study multiple targets. Proteochemometrics draws heavily on existing QSAR techniques, such as using topological indices or physicochemical properties as molecular descriptors or using PLS for model building, which the approach requires; however, it has spurred the development of complementary techniques to characterize the amino acids comprising protein targets and the differential selectivity profiles of molecules and targets. This is an important area for future research, especially if it proves possible to encompass nonhomologous proteins and structurally diverse sets of ligands. Techniques such as QSAR and proteochemometrics focus on the details of ligand–protein interactions. Systems biology, as its name implies, sets itself the far more challenging task of computationally modeling entire living organisms. Within this broad area, Patterson argues in Chapter 4 that the dependence of a drug's clinical effect on its chemical structure (broadly defined) is currently best analyzed using the gene expression spectra of cells that have been exposed to bioactive molecules, and describes some initial attempts to apply current QSAR techniques to the analysis of such data. The results are far from conclusive, but they do suffice to demonstrate the potential of such approaches, which are likely to become of increasing importance as more expression spectra become available in the future.

A vital factor in any attempt to relate chemistry and biology is the way in which the molecules are characterized for modeling. Both the structure and the properties of a molecule can be regarded as different manifestations of the same underlying wave equation, and it is thus to be expected that molecular descriptors will indeed be related to property; however, the extent of the relationship will depend on the descriptors, the properties, and the sets of molecules, *inter alia*. Sukumar et al. in Chapter 5 focus on the first of these, providing a wide-ranging review of descriptors that can be generated not just for small molecules but also for their biological targets. The review covers both well-established descriptors, such as physicochemical properties and topological indices, and some more recent and/or more complex descriptors, such as topomers, bioactivity spectra, and Fourier surface transforms. Descriptor selection is a challenging task, not least because there is often a trade-off between interpretability and predictive power; there will thus be a continuing need for new descriptors that can reconcile these two conflicting criteria. In Chapter 6 Misra et al. present one such novel descriptor, the chemical signature, that characterizes the immediate neighborhood of each atom in a molecule. Such environmental descriptors have been known since the early days of chemoinformatics: What is novel here is that the same basic technique can be used to describe not just atoms but also bonds, entire molecules, and reactions; and the authors describe experiments in which the atom-level signatures are used for QSAR and QSPR. By analogy with atom-level signatures, it is possible to derive signatures describing the neighborhoods of amino acids in proteins; the use of these is illustrated by their use for the prediction of protein–protein interactions; and the use of both types of signature in combination is illustrated by studies of enzyme–metabolite and drug–target interactions.

Once the chemical and biological entities of interest in a study have been given machine-readable representations, one must be able to link those representations together for the purposes of modeling and prediction. This is being done increasingly by means of workflow, or pipelining, tools, which have rapidly established themselves as a simple, highly effective way of integrating and analyzing heterogeneous data sources. In Chapter 7 Meinl et al. prove an overview of the four tools that have been used most extensively with chemical and biological data sets: InforSense, KNIME,

Pipeline Pilot, and Taverna. Although there are differences between these tools (e.g., in their licensing status, the ways that workflows are constructed, and the manner in which sets or subsets of data are processed), they share a common basic model, that of a directed acyclic graph, through which data and intermediate results flow to yield the final output of an analysis. The operation of such tools is illustrated by the use of KNIME for virtual high-throughput screening, the analysis of cell images, and text mining in PubMed abstracts.

Advances in the technologies of sequencing and structure determination mean that there are now large numbers of proteins for which the function is unknown, and this has spurred the development of computational methods that can suggest the function of proteins for which this information is unavailable. Favia and Nobeli in Chapter 8 describe the use of docking for this identification task. Ligand–protein docking has been discussed extensively in the literature as an effective tool for virtual screening when the 3D structure of the biological target is available. Function prediction involves docking known substrates into a protein as a way of suggesting the possible function (or functions) of that protein. The authors highlight some of the differences between this application and conventional structure-based virtual screening, focusing in particular on the use of the intermediates of plausible chemical reactions to identify the catalytic functions of enzymes and the evidence that is now available supporting the view that proteins are functionally promiscuous. Finally, in Chapter 9 Nigsch considers not just individual proteins but the ensemble of proteins and molecules involved in the set of biochemical reactions that comprise a well-defined function in a cell. Such a biological pathway is highly complex; but it forms just one part of the biological network that describes all of a cell's processes. Methods are now under development that may enable small molecules to be used both to elucidate and to affect the functioning of networks: as the author notes, this work could form the basis for the rational modulation of diseased cells. The introduction of such novel health therapies is the aim, either implicit or explicit, of all the work described in this book, and it will thus be most interesting to see how the techniques suggested here develop over the next few years.

REFERENCES

1. Ray, L. C.; Kirsch, R. A. Finding chemical records by digital computers. *Science* **1957**, 126, 814–819.
2. Hansch, C.; Maloney, P. P.; Fujita, T.; Muir, R. M. Correlation of biological activity of phenoxyacetic acids with Hammett substituent constants and partition coefficients. *Nature* **1962**, 194, 178–180.

3. Willett, P. From Chemical documentation to chemoinformatics: fifty years of chemical information science. *J. Inf. Sci.* **2008**, 34, 477–499.
4. Warr, W. A. Balancing the needs of the recruiters and the aims of the educators. Presented at the 218th American Chemical Society National Meeting, New Orleans, LA, Aug. 22–26, 1999. http://www.warr.com/warrzone2000.html.
5. Paolini, G. V.; Shapland, R. H. B.; van Hoorn, W. P.; Mason, J. S.; Hopkins, A. L. Global mapping of pharmacological space. *Nat. Biotechnol.* **2006**, 24, 805–815.

PETER WILLETT

University of Sheffield

PREFACE

Despite their similar names, in many ways cheminformatics and bioinformatics address rather different problems. Much of bioinformatics focuses on the analysis of sequences, and even when addressing chemical structure, structural biology generally tend to address larger biomolecules (i.e., proteins). Cheminformatics, on the other hand, tells us how to handle smaller molecules, and it is more closely tied to typical chemical problems such as structure representation and reaction modeling. Further differences emerge in a variety of areas, such as software tools, databases, and algorithms employed. Yet the two fields also exhibit many commonalities, such that many cheminformatics approaches can be applied successfully to bioinformatics problems (and vice versa). Given that many computational methodologies, such as machine learning and graph algorithms, are employed in both fields, this is not surprising—and this book aims to extend the synergistic relation between both communities even further, by stressing the commonalities in science, both those already explored as well as more novel ones.

Our concrete motivation for embarking on this project was the observation that even though many bioinformatics applications in one way or another do consider chemical structures, in most cases this was on a rather superficial level. Conversely, much of current cheminformatics work focuses purely on chemical structures alone, thus ignoring the biological context of a molecule's behavior. It appeared to the editors of the current book that this situation should be improved upon. To bring together scientists working at the intersection of the cheminformatics and bioinformatics fields, one of us (R.G.) organized an ACS symposium (*Cheminformatics Techniques in Bioinformatics*) to highlight examples of such interdisciplinary applications. The breadth of topics discussed at the symposium, as well as the interest shown by speakers and audience alike, suggested that a collection of articles specifically focusing on how cheminformatics methods are employed in bioinformatics scenarios would be well received. Hence, we solicited contributions from speakers at the symposium as well as from scientists in the wider field.

Although the final book has been long in coming, we feel that the wait has been worthwhile. Owing to our contributing authors, we have been able to put together a broad collection that covers data sources, methodologies, and applications, all highlighting the intersection of the cheminformatics and bioinformatics fields. This has happened in a manner that will be widely accessible to readers, ranging from university seniors and graduate students to practitioners working in the pharmaceutical industry and related fields. Our primary hope from this book is that readers will gain a broad view of how cheminformatics techniques are applied in bioinformatics settings, and more generally, how information about small-molecule structure can be integrated with larger biological systems (in particular, those modeled in the computer, in silico). With the advent of chemical biology, systems biology, and related areas of study, the computational problems in these fields will without doubt necessitate a combination of bioinformatics and cheminformatics techniques. To bring it to a single point, one can say that *sequences as strings are handy, but it's the molecules that are really doing the work!*

We would like to note that this book would not have come about if not for the support and perseverance of Anita Lekhwani from Wiley. I (R.G.) would also like to thank Leah Solla for supporting my efforts in organizing the ACS CINF symposium that initiated this effort, and we would of course like to thank all the contributing authors whose efforts and patience contributed to the final contents of the book.

Rajarshi Guha
Andreas Bender

CHAPTER 1

BRIDGING CHEMICAL AND BIOLOGICAL INFORMATION: PUBLIC KNOWLEDGE SPACES

PAUL A. THIESSEN
National Center for Biotechnology Information, National Library of Medicine, National Institutes of Health, Bethesda, Maryland

WOLF-D. IHLENFELDT
Xemistry GmbH, Lahntal, Germany

EVAN E. BOLTON and STEPHEN H. BRYANT
National Center for Biotechnology Information, National Library of Medicine, National Institutes of Health, Bethesda, Maryland

1 LANDSCAPE OF PUBLIC CHEMICAL (BIOACTIVITY) DATABASES BEFORE PUBCHEM

At the time of this writing, PubChem[1] is probably the most widely known publicly accessible chemical compound database on the World Wide Web (WWW, or just Web). It contains not only chemical structures, but also biological data linked to these structures. PubChem was launched in 2004, but it is certainly not the first freely available, Web-accessible database providing biological information on the Internet.

The biological data landscape is complicated by varying definitions of what classes of information should be considered as biological information. Do toxicity data constitute biological information? If yes, should a qualifying database contain actual measurements, or can this information be provided in distilled, abstracted formats, perhaps even as material safety data sheets (MSDSs) or simple handling classifiers? Do we simply

Computational Approaches in Cheminformatics and Bioinformatics, First Edition.
Edited by Rajarshi Guha and Andreas Bender.

consider biological information in the context of drug research, or is basic biological data (e.g., metabolic pathways) part of the picture?

The following descriptions of databases launched before PubChem should not be considered comprehensive but, rather, an editorially selected collection, highlighting novel features and the influence that these systems had on the development of later systems. Several sites have attempted to catalog all major Web-accessible chemistry databases (e.g., the Chembiogrid[2] resource), which the reader may want to consult for a broader picture. Additionally, an overview of chemistry and the Web in 1998 was published in a special issue of *Chimia*.[3]

The Protein Data Bank[4] (PDB), begun in the 1970s and available on the Web since the early 1990s, can be considered a grandfather of chemical structure databases, although with a rather peculiar and narrow focus. PDB stores and redistributes crystal structures of proteins and other biological macromolecules. This includes proteins with bound small molecules, information of high biological relevance. The actual structures have always been available for download, from basic FTP sites, shipped tapes or CDs, or the current Web interface. Nevertheless, small molecules and bioactivity data were never the principal focus of this database. Even today, the extraction of small ligand molecules from available data files remains a challenge, due to the particularly limited and often abused encoding standards employed. Only recently has PDB begun to provide nontextual ligand search capabilities. Link-outs to biological activities stored in external databases are still absent. PDB has stood the test of time and provides unique information but is rather isolated on the Web, despite numerous databases making the effort to establish relationships between PDB entries and their data (via unidirectional links).

Among the original small-structure chemistry databases making an entrance during the dawn of the Web, ChemFinder[5] by CambridgeSoft (development started in 1995) was probably the most influential and most professionally managed system. This was not, however, the first widely recognized small-molecule repository—that honor probably goes to the NIST WebBook[6] (online since 1996), but it contained only nonbiological data such as spectra and physical constants. ChemFinder pioneered many of the query and interface techniques still used today in Web chemistry databases, such as intelligent query parsing, structure search capabilities, and link-outs to secondary databases. Like PubChem (more details to follow), ChemFinder did not attempt to store all information located but, rather, linked to the original source. Because CambridgeSoft is the developer of the widely used chemical structure drawing program ChemDraw, ChemFinder was also designed as the showcase for the Web browser plug-in variant of ChemDraw. Using the ChemDraw plug-in, this was the first database to provide comfortable interactive drawing of structures for full-structure and substructure queries on the database, although at the expense of using a nonportable Microsoft Windows/Netscape-only interface (at the time of launch). Originally, ChemFinder was not specifically concerned about biological activity links. It indexed sites that the development team deemed important and indexible with the technologies available to the engineers, which included rather sophisticated chemistry-aware text-matching algorithms, allowing the establishment of database links even in the face of spelling variants and misspellings. The original ChemFinder database is no longer accessible. CambridgeSoft is relaunching it under the ChemBioFinder brand. The new release directly incorporates various drug databases, such as the *Merck Index*[7] and the National Cancer Institute (NCI) Developmental Therapeutics Program (DTP/NCI) cancer and antiviral screening data.[8]

The DTP/NCI database contents were prominent in the history of bringing biological data to the Web. This data set was first available on the Web via the NCI database browser[9] (currently in version 2). The first version of the NCI database browser was released in 1998, with about a quarter of a million structures from DTP/NCI. This compound set was collected over four decades but had only been accessible by an in-house system at NCI. The biological aspect of the database included the results of tumor cell line screenings of these compounds, measured on a collection of standard tumor cell lines. A smaller subset of compounds was also subjected to antiviral screens, with a special focus on anti-AIDS activity. The original compound data was (and is) problematic—many structures were registered without stereochemistry, and even the reconstruction of the connectivity of some structures is not always possible in an unambiguous way, due to the original coding of the in-house registration system.

The NCI database browser pioneered many important features. Among the Web structure databases of the time, it had the most sophisticated query system (even by today's standards), including features and abilities such as dynamically generated query forms (via JavaScript) and advanced tools to merge, manage, and store query and hit lists. Another important functionality in the design of this database was extensive export options for result sets with dynamic format conversion, enabling the use and reuse of the database contents for local projects. Until the advent of PubChem, this functionality was largely overlooked, with Web interfaces to public resources (even to this day!) designed with the single purpose of human browsing, with meager export capabilities—only parts of the records or a single full record at a time. Restrictive public resources with insufficient data filtering and export capabilities make the goal of reusing and reanalyzing public datasets very difficult to realize.

The NCI browser was among the first major chemistry database systems on the Web to implement a platform-neutral interface for structure searching and three-dimensional (3D) visualization. For structure input, it relied on the (then) newly released JME Java structure drawing applet,[10] an important development and a popular tool even today. Its result-display routines pioneered the use of dynamically generated GIF images of structures, where a query was displayed directly on the results using structure highlighting and other annotations depending on the query, a rarely found feature even now. For 3D visualization, the browser was first to support the export of structure models as virtual reality modeling language (VRML) files—at the time a highly promising general 3D display standard for the Web, but no longer well supported in the Web ecosphere. More common in chemical applications now are Java-based approaches such as JMol[11] for 3D chemical visualization. While the use of platform-independent approaches for public Web systems is now considered mainstream, at the time there was considerable dependence on external helper applications (e.g., RasMol[12]) and platform-specific plug-ins (e.g., MDL Chime[13]).

While the NCI database browser was a pioneer for the distribution of assay data, the Klotho database[14] (now defunct) was similar in that it was the first system to link biological pathway data with small molecules. Although not a direct successor, KEGG[15] (started in 2000) is now assuming its role. KEGG's PATHWAY database provides information about the role of small molecules in biological pathways, while the LIGAND database and its various sub-databases summarize data on chemical structures in the KEGG collection. A unique feature of KEGG is that it contains reaction information, linking the transformation of structures, although without an exact atom mapping (which the commercial database Biopath[16] has). Additional important databases in the biological pathway context are the Human Metabolome Database[17] (HMDB, online since 2004) and the BRaunschweig ENzyme DAtabase[18] (BRENDA, online since 2003).

PubChem is not the first public chemistry database supported by a long-term U.S. government sponsoring commitment. ChemIDplus,[19] which, like PubChem, is maintained under the umbrella of the National Library of Medicine (NLM), is older than PubChem. This database is important because it is considered one of the most extensive public toxicological information resources on the Internet. ChemIDplus contains nearly 400,000 records, many containing detailed toxicity information, and is linked to the NLM Toxicology Data Network (TOXNET) cluster of related databases, such as TOXLINE[20] (toxicology literature), GENETOX[21] (genetic toxicology), and CCRIS[22] (carcinogenicity and mutagenicity data).

A database that can be considered a direct precursor of PubChem is ChemBank,[23] launched in 2003 with support from the NCI Initiative for Chemical Genomics (ICG). ChemBank contains nearly 1.7 million chemical structures generated from vendor catalogs and filtered by various computational criteria relevant for drug design (e.g., rule-of-five[24] compliance, substructure and element exclusion, drug likeness score, and chemical diversity). These compounds are linked directly to biological screening results. To our knowledge, ChemBank is the first public chemical bioactivity database service supporting a dedicated application programming interface (API) for remote programmatic queries by applications other than Web browsers, through a set of Simple Object Access Protocol (SOAP) functions. This is similar to the more extensive interface that PubChem provides with its Power User Gateway[25] (PUG).

The landscape of publicly accessible databases with chemical and biological content has expanded since the launch of PubChem. Some databases involving sizable data curation efforts, such as ChEBI[26] (2005) and DrugBank[27] (2006), focus on drug and drug candidate information. Some databases involving sizable biological activity literature abstraction efforts, including BindingDB[28] (2001), PDB-Bind[29] (2004), KiBank[30] (2004), and BindingMOAD[31] (2005), focus on small-molecule binding constants. The earliest of these, BindingDB, is notable in that it allowed research groups to contribute data directly. The effective takeover by the European Bioinformatics institute (EBI) of a major commercial bioactivity knowledge base—the BioFocus DPI StARLITe[32] database—is a recent development that may be considered nothing short of remarkable given the breadth and depth of bioactivity information directly relevant to drug discovery. The StARLITe database, integrated into the public knowledge bases at EBI (e.g., ChEBI), will be a welcome addition to the publicly accessible space, with more than 2 million bioactivity data points abstracted from 12 journals for about 1500 drugs, 10,000 drug candidates, and 450,000 drug leads. This may be a sign of things to come as public knowledge bases grow in size and quality, potentially limiting the space in which commercial vendors can operate.

2 PUBLIC DATABASE INTEGRATION EFFORTS

Unlike the databases mentioned above, PubChem is neither an originator of bioactivity information (e.g., DTP/NCI or ChemBank) nor is it a curation or literature abstraction effort (e.g., ChEBI, KEGG, or PDB-Bind). It has no tiered data access scheme, no log-in requirement, and no restriction on who may contribute. PubChem is an open repository, depending entirely on external contributors for its content. PubChem was originally funded as a part of the Molecular Libraries Program (MLP), a component of the National institutes of Health (NIH) Roadmap.[33] This program includes the Molecular Libraries Probe Production Center Network (MLPCN), consisting of grant-supported experimental laboratories, and a shared compound repository referred to as the Molecular Libraries Small Molecular Repository (MLSMR), offering biomedical researchers access to chemical samples. The MLPCN is the successor to the Molecular Libraries Screening Center Network (MLSCN) from the initial MLP pilot phase.

PubChem archives the molecular structure, bioassay data, and annotations from the MLP and third-party depositors. PubChem provides search, retrieval, and data analysis tools to optimize the utility of information collected. Also, PubChem imports other public sources of chemical structure and bioactivity information, and integrates this with data contributed and NIH biomedical knowledge bases (e.g., PubMed,[34] MMDB,[35] GenBank,[36] MeSH,[37] DailyMed[38]). The primary aim of PubChem is to provide a public online resource of comprehensive information on the biological activities of small molecules, accessible to molecular biologists and to computational and medicinal chemists.

ChemBank, perhaps the public data system closest in content and form to PubChem, is similar in various ways. Both databases contain large amounts of small-molecule structures and associated bioassay data. Both databases provide tools to search and analyze these data. Both databases have similar stated goals of providing freely available information. There are considerable contrasts, as well. ChemBank is not an open repository. ChemBank data are generated locally at the Broad Institute, giving ChemBank complete control over the content and verbosity (i.e., the "rawness") of biological screening data provided. PubChem takes chemical structure, screening information, and other data from many organizations, including ChemBank, each with its own ideas on what is necessary to communicate experimental results to the public. ChemBank embargoes all new data from public access as a matter of policy. As such, ChemBank can be accessed either in public form or via the data-sharing agreement (DSA), which grants access to both public and embargoed data. PubChem normally releases data immediately once the depositor is satisfied with the data import accuracy and overall presentation within PubChem (using the PubChem Deposition Gateway[39]). It is possible for a PubChem depositor to put data on hold: for example, to synchronize the release of PubChem records with the publication of a paper or announcement of a new resource, but PubChem does not hold data as a matter of policy and there is no way for any user (even the originating depositor) to search or analyze such data, as with the ChemBank DSA. Considering that PubChem is run by a government agency, it is restricted from tracking public data users so as not to violate privacy laws, precluding the use of any tiered or collaborative data access model requiring a login, such as ChemBank DSA.

There are strengths and weaknesses in both these database models. PubChem's open model prevents its use for selective release prior to publication, whereas selective release is often the preferred method in the highly competitive scientific "publish or perish," intellectual property–centric environment that exists in both the private and academic sectors; but PubChem lets anyone integrate and cross-link their own data with those of many other data originators and biomedical knowledge bases with minimal effort, as most of this analysis is performed automatically upon deposition. The ChemBank collaborative model provides freedom to decide who gets to see what and when, but it is restricted to data originated by ChemBank collaborators. Although many other parallels and contrasts can be made, it is really the data collection policy that sets PubChem apart from ChemBank and other databases. Given its unique nature as a freely available, public, and open archive, future discussion will focus primarily on PubChem: its contents, methods of integrating data from disparate sources, and the caveats involved in such a system.

3 DESCRIPTION OF DATA CONTENTS

There are two primary aspects of a chemical database: the policy and procedures by which the database is populated, and the actual data it contains. Some are curated manually; that is, records are entered and checked by a human (e.g., KEGG or BioCyc). On the opposite end of this spectrum is PubChem, whose data entry is entirely automated, with data provided by depositors treated as is. ChemSpider[40] is between the two, where much data acquisition is automated, but individual records are open to manual adjustment. One may legitimately argue the value of the various approaches, but the biggest factor here is a combination of the size of the database and the human-hours available for manual data entry and validation, through either direct staffing or a wiki-style open system.

PubChem collects information from depositors via the PubChem Deposition Gateway largely to the degree of detail they are willing to provide. For substances, the only field required is an external registry ID; however, a rich set of information may be and often is provided, including a chemical structure in Structure-Data File[41] (SDF), the Simplified Molecular Input Line Entry System[42] (SMILES), or IUPAC International Chemical Identifier[43] (InChI) format; uniform resource locators (URLs) to external Web-enabled resources; substance synonyms; comments; and annotations in the form of cross references to NCBI knowledge bases (including PubMed, GenBank, MMDB, Gene, Taxonomy, OMIM, and Probe). Many of the databases mentioned earlier are also contributors to PubChem, including BindingDB, ChEBI, ChemBank, ChemIDplus, DTP/NCI, KEGG, and MMDB. Each PubChem depositor may classify the types of information they provide (e.g., biological properties, substance vendor, chemical reactions). Although the contribution to PubChem by individual depositors varies, a deposition often serves to inform PubChem users that additional information is available at the depositor's website.

Visited on average by more than 70,000 unique users per day, PubChem contains (at the time of writing) over 85 million substance descriptions from over 160 depositors, 500,000 bioassays with more than 140 million substance outcomes originating from over 45 depositors, and more than 30 million unique chemical compounds. A current list of PubChem contributors and record counts is available.[44] The data content is still growing, especially with respect to assays and assay test outcomes; the number of unique chemical compounds grows more slowly as PubChem's coverage of known chemicals increases.

PubChem is organized as three distinct databases: substance, compound, and bioassay. The substance database contains depositor sample descriptions, necessarily including any chemicals tested in any assays by that depositor. Compound records are created from PubChem substances through the means of standardization or normalization of deposited chemical structure information. (Standardization is described in more detail later.) As such, the PubChem compound database is derived from the PubChem substance database and represents the overall unique chemical structure content of PubChem. Each PubChem compound record consists of a fully defined chemical structure (no implicitly defined valence, hydrogens, or aromatic bond definitions), computed properties (e.g., molecular weight, molecular formula, hydrogen-bond donor counts), and identifiers (e.g., IUPAC name, SMILES, InChI).

PubChem bioassay records are more complex than substance definitions. They consist of two logical parts, a definition and associated data. Unlike PubChem substance records that have minimal requirements, bioassays require a more complete record from a depositor, with separate sections for description, protocol, target definition, comments, readout definitions, URLs, and annotations. After a bioassay definition is created by a depositor, data for the assay may be given for each substance tested in the bioassay, including an activity outcome (for partitioning purposes, e.g., active, inactive, inconclusive), activity score (for sorting purposes, e.g., a value between 0 and 100, with larger values indicating more active or more interesting results), URL, and annotations. A PubChem bioassay record may be configured to provide "panel-style" results, such as in the case of phenotypic assays or selectivity profiling assays where there are multiple outcomes. These panel assays may have multiple sets of targets, outcomes, and results defined, providing for a compact and data-rich representation.

Bioassays may contain a target definition, typically a specific protein. These allow (indirect) cross-links to be formed between GenBank entries and PubChem substances and compounds. Thus one may group proteins of interest (e.g., with sequence analysis tools such as BLAST[45]) and thereby discover sequence families whose members have been targets of PubChem assays and chemicals that have been found to interact with these proteins. Assay targets may also be nucleotides and may have cross-links to genes and taxonomy. PubChem precomputes similarity between known assay targets, allowing one to cluster or navigate by target similarity. One may also search PubChem assay targets by sequence, using a specialized subset of the online NCBI BLAST resource.[46]

Another important source of biological information for PubChem records is the NLM Medical Subject Heading (MeSH) classification. This ontology of controlled vocabulary is used to index articles in the biomedical literature, containing many chemical names and their known biological function. For example, the term *aspirin* has a description, a list of synonyms (acetylsalicylic acid, acylpyrin, etc.), pharmacological actions (anti-inflammatory agents, nonsteroidal, fibrinolytic agents, etc.), a place in a tree of chemical structure classification (salicylic acids), and categorized links to PubMed journal articles that refer to aspirin. PubChem automatically matches the names supplied by depositors to the MeSH terms and links to MeSH and PubMed as appropriate. The reciprocal links are also present; that is, through a simple link in Entrez, one may easily get a list of all PubChem substances or compounds that share the MeSH pharmacological action "fibrinolytic agents," which includes the record for aspirin. Similarly, one may find PubMed articles that refer to a particular chemical, or conversely, list chemicals referred to in an article. PubChem thus uses MeSH as both an annotation and classification system, and as a direct link between chemicals and the biological literature. Although not perfect—not all chemicals mentioned in an article are indexed, and not all articles in PubMed are MeSH-indexed—integration of PubChem and MeSH provides a valuable tool to learn about the biological function of small molecules and the literature available.

As mentioned above, the Protein Data Bank (PDB) is a direct source of detail on small molecules' three-dimensional interaction with proteins.

NCBI maintains an effective mirror of PDB called the molecular modeling database (MMDB), from which PubChem extracts the small molecules (ligands). This process is fully automatic, and some details and problems with that are discussed more fully below. MMDB ligands provide a direct link between PubChem records and Entrez's structure database, and thence to proteins and nucleotides in GenBank.

Links to a wide variety of other types of biological data are indirectly available through PubChem in records imported from other public databases. For example, one may find information on drugs and metabolic pathways (e.g., KEGG and BioCyc[47]), toxicology (e.g., ChemID Plus, EPA/DSSTox[48]), cancer screening (e.g., DTP/NCI[49]), anti-HIV screening (e.g., DTP/NCI, NIAID[50]), imaging agents (e.g., MICAD[51]), scientific journals (e.g., *Nature Chemical Biology*[52]), NMR data (e.g., NMRShiftDB[53]), and protein-binding affinity (e.g., BindingDB), just to name a few.

4 TECHNICAL ASPECTS OF BIOACTIVITY KNOWLEDGE BASE DATA INTEGRATION

In many ways, collection of information from data contributors is the easy part. There are many aspects to integration of such data to maximize navigation and interpretation. Although not meant to be an exhaustive discussion on the topic, describing concepts and ways in which this is achieved within the scope of PubChem is the primary goal of this section.

5 CHEMICAL STRUCTURE HANDLING

Chemical structure is often the key to joining disparate data sources. Each depositor can have completely different textual names which they use to refer to a substance but still have the same chemical structure. A simple text search of these two records would not reveal that they are related. If the chemical structure is the same, the list of textual names may be combined along with any other information known about the two substances. This is the power of the chemical structure.

A primary step toward using chemical structure in an integrated biochemical resource is determining when two or more records are actually referring to the same chemical, so that the records may be linked together with that chemical as the commonality. To do this correctly requires knowing in full detail the exact chemical structure being represented. This can be problematic when integrating multiple heterogeneous sources, all of which may have their own means of describing small molecules, in a greater or lesser level of detail. In this section we present some of the difficulties in comparing chemical species provided by disparate depositors and the concepts behind the tools that PubChem uses to provide effective integration and cross-linking of chemical structure information.

Standardization of Chemical Structure Representation

There are no universally adopted rules on how to convey chemical structure data. As such, each organization (or chemist within an organization) is free to adopt chemical drawing conventions arbitrarily. PubChem receives data from many different organizations, compounding the issue, as such drawing conventions may conflict between depositors. Specialized processing is required to normalize the representation of chemicals in a consistent way to allow data from different depositors to be combined and integrated when the two records are associated with the same chemical structure. Such standardization processing requires modification of the original data provided by the depositor.

Making such changes directly to the chemical structures at the time of import into PubChem would violate the very archival nature of PubChem, intent on preserving data provided by depositors. Furthermore, it may be rather unsettling to the contributing organization to have their data modified, especially if there is a mistake in such processing. To balance the competing demands of preserving original information and the need to unify chemical structure representation, to the best of its ability, PubChem retains the deposited chemical structure information and, after successful normalization processing, associates one or more PubChem compound records with the substance record. This allows PubChem processing rules to change as a function of time while preserving the integrity of the original deposited data.

Actual chemical structure processing steps used by PubChem involve a series of verification steps, including atomic element checks, functional group normalization, and atom valence checks; standardization steps, including valence-bond representation normalization (for tautomer-form invariance), aromaticity normalization (for VB-form invariance), sp^2/sp^3 stereochemistry detection (including systems involving allenes, allene ring equivalents, or free electron pairs), and explicit hydrogen assignment; and mixture component processing steps, including covalent unit detection, proton-based neutralization (when applicable), and parent assignment. Components detected during processing are individually reprocessed in the same way. The final processed structure and unique components are associated with the substance record as standardized compounds. If a substance fails normalization processing, no PubChem Compound record is associated with the substance. PubChem provides a standardization service[54] to allow users to standardize chemical structures: for example, to determine what PubChem Compound record(s), if any, are associated with these chemical structures, an especially important step when seeking to integrate private data with PubChem.

Identity Groups

One of the compelling reasons for creating a chemical biology resource consisting of various depositors is the ability to compare bioactivity data between depositors. Standardization of chemical structures into a uniform representation is a big first step in this direction, but it does not resolve the issue completely. Different depositors may provide variable levels of chemical structure detail of what many would consider to be essentially the same structure, depending on purpose. For example, two depositors may have the same chemical structure as far as connectivity is concerned, but one depositor may have provided a complete stereochemical description whereas the other did not. An additional concept employed by PubChem to locate related structures is "identity groups" to provide variable degrees of "sameness" at the level of connectivity, where chemical structure connectivity (atoms/bonds) is identical but variation may occur at the isotopic and stereochemical levels; stereochemistry, where chemical structure connectivity and stereochemical centers are identical but different isotopic forms are allowed; isotopes, where chemical structure connectivity and isotopic form are identical but variation is allowed in stereochemical centers; exact, where chemical structures must be identical at all levels (connectivity, stereochemical, and isotopic); and any tautomer, where a more liberal tautomeric identity representation is used that considers tautomerism that occurs under mild (acidic, basic, or temperature) conditions, allowing for variability at the connectivity, stereochemical, and isotopic levels. (Internally, these groups are implemented as constant-width 64-bit structure hash codes,[55] which are compared easily and efficiently and are faster to compute than canonic linear structure representations.)

Mixtures and Parents

Chemical structures in PubChem are often provided in various forms, salts with different counterions, formally charged, neutralized, and so on. PubChem standardization detects mixture components and associates these components with the record for the entire mixture. For example, the chemical structure of sodium acetate will be linked to the components sodium and acetic acid. Thus, one may "expand" or "collapse" a list of chemical structures to explore mixtures or components of compounds in which one is interested. When considering biological activity, it often does not matter in what form a compound is tested (salt or nonsalt), such that it may be useful for bioactivity comparison purposes to consider acetic acid and sodium acetate as being the same structure. To handle this effectively, PubChem uses the concept of a parent compound. In the case of sodium acetate and acetic acid, acetic acid is assigned as the parent. Not every compound will have a parent. Complex mixtures containing similarly sized organic moieties, or purely inorganic mixtures, do not have parents because here the concept of "parent" has ambiguous meaning. The notion of the parent compound allows a compound search to be expanded to a larger number of chemical structures that are presumably biologically equivalent; simultaneously, bioactivity data may be collapsed to a smaller number of equivalent compounds.

Similar Compounds

The combination of the paradigms of parent compound, identity group, and standardization provide very powerful means to navigate bioactivity information of "same" chemical structures. PubChem provides similarity neighboring relationships for structures that are not the same but are very similar: for example, analogs in a Markush-type series. This allows a user to expand the scope of compounds or substances considered to include additional chemical records that are structurally similar and thus that may have similar biological activity. These neighboring relationships, precomputed for each compound record, are equivalent to a PubChem 2D similarity search using the Structure Search tool[56] at a 90% threshold.

6 STRUCTURE SEARCHING

PubChem's compound database is searchable by all standard 2D structure search methods: full structure (internally performed via hash code comparison), substructure and superstructure, structural similarity (using the classical approach with screening bit vectors for acceleration and similarity score computation), and chemical formulas. The similarity comparison uses a special boosting scheme to assign scores above 100% for identities not normally distinguished by this algorithm: 104% for full isotope and stereo identity, 103% for either stereo or isotope match but not both, 102% for connectivity identity, and 101% for tautomer identity.

7 TEXT SEARCHING

In this section we outline some of the problems associated with plain text searching in a chemical structure database. There is a notorious variation in how molecules are named, even among knowledgeable chemists; and, of course, there is always the possibility of error. All of these problems are multiplied in a large chemical database with redundant structures, and even more so in a database such as PubChem, which brings in data from numerous sources. Here we describe some weighting strategies used by PubChem to increase the likelihood that a text name search will result in the correct structure.

Search Fields and Errors

An immediate problem is one of interface. If there is a single text box for user data entry, the vast majority of users will simply enter the name there without understanding the underlying search details. For example, PubChem's entry for acetaminophen (Tylenol, CID 1983) has the name "aspirin-free Anacin" as one of the synonyms supplied. Hence, a completely unrestricted text search in PubChem for the term "aspirin" will bring up the record for acetaminophen. It is possible to avoid this to some degree by narrowing the search appropriately using the "CompleteSynonym" index; a search for "Aspirin[CompleteSynonym]" in PubChem will be more reasonable but still finds cases where the record is a mixture of multiple chemicals. Take, for example, CID 24847967, where the depositor has misused the synonym field when submitting to PubChem and has specified the names of the chemical components of the mixture—aspirin and oxycodone hydrochloride—as separate names for the mixture itself.

As the complexity of the molecule increases, so does the chance that different sources will disagree on—or simply make a mistake about—the details of the structure. Stereochemistry is a common example of this sort of problem; searching "vancomycin" in PubChem's Compound database (again without restriction) currently yields 71 structures; if narrowed to synonyms that are exactly "vancomycin," the search still yields six structures. The correct structure (CID 14969, according to ChemSpider) has 28 neighbors with the same atoms and bonds but different stereochemistry, four of which are called "vancomycin." Even for a human chemist, determining what the correct structure is may be problematic, and for a computer alone, nearly impossible.

Weighting

With all of these problems, one might ask: How can a PubChem search for "aspirin" show the correct structure of aspirin as the first result? (Keeping in mind that unlike many other databases, PubChem has no mechanism for a curator or expert to step in and identify one manually.) The answer is that PubChem attempts to weight particular terms more heavily than others, and sorts the records based on the weight of the terms that are matched to the original query. This is not a novel strategy, but the important part is how the weights are determined.

Here PubChem can actually take advantage of its diversity of information sources using a voting scheme to weight common names more heavily. That is, if many depositors supply the name "aspirin" for a given chemical structure but only a few supply the name "acetosal" for that same structure, the name "aspirin" gets a higher weight and a search for "aspirin" brings to the top the record for which the most depositors gave that name. PubChem gives a name only one "vote" per source, not one vote per source per record; this prevents sources that provide many records of the same structure from falsely imbuing too high a weight, such as MMDB, which contains hundreds of heme ligands called "Hem"—not a highly informative name.

PubChem also weights a name based on how many upper- and lowercase letters, numbers, and symbols (dashes, primes, etc.) appear in the name. The exact formula is not terribly important and is subject to change and so is not given here, but the effect is that more readable names, such as "aspirin," get higher weights, while short, long, or numeric names get lower weights (e.g., "Sine-Off Sinus Medicine Tablets-Aspirin Formula").

8 IMPORTING DATA FROM NONCHEMISTRY SOURCES

Some public scientific knowledge bases (e.g., PubMed) are devoid of well-defined chemical structure content (i.e., atom or bond description), containing only references to chemical names. In some cases, these resources have a focus other than chemistry (e.g., biological macromolecules, pathways, diseases). To bridge this gap between these knowledge bases, when detailed chemical structure information is simply unavailable, solutions must be created. Chemical name matching is an obvious choice: simply looking up the chemical name and, if there is an exact match, assigning the chemical structure to that data record; however, this problem is far from solved and has many caveats.

Deriving Chemical Structure from Chemical Names

Deriving a chemical structure from a chemical name is nontrivial. A single chemical structure can have many chemical names. For example, in PubChem, aspirin (CID 2244) has more than 300 synonyms, being a mix of IUPAC or CAS names (of different generations), common names (some in different languages), product names, various registry names, and so on. A chemical name can also match multiple records. Again, using PubChem, an exact search for "aspirin" currently returns 15 results; one result is aspirin, another is an isotopic-labeled form of aspirin, and the rest are mixtures where aspirin is a component. Although a match may be found to a chemical name, that match may not be desired. One of the 15 exact matches to "aspirin" in the example above, mentioned previously, consists of a mixture of aspirin and oxycodone. It is also possible that the particular chemical name is not found even though the structure represented by the synonym may be available. These caveats attempt to stress how simple text matching of chemical names may readily provide false positives and false negatives that could be difficult to address in an automated fashion.

One may attempt to construct some controlled vocabulary for name matching, where a list of authoritative names are linked to a known structure. This is the method used by NCBI's MeSH database, where expert analysts match references to chemicals in the biomedical literature to particular entries in the MeSH database. CAS registry numbers[57]are also used for this purpose, although this is not as straightforward as it appears to be, because frequently, multiple CAS numbers have been assigned: to stereoisomers, isotope label variants, mixtures, formulations, extracts, and so on.

To provide a bridge between the text and chemical structure worlds when depositions are text only, PubChem provides the ability to generate chemical structures from chemical names at deposition time in three primary ways: when a synonym is a PubChem Compound identifier (e.g., CID2244); via MeSH, when a synonym matches one found in a MeSH record and when PubChem assigns just a single PubChem Compound record to the MeSH record; and using name-to-structure software (LexiChem[58]). Chemical structures generated in this fashion are not considered to be a part of the deposit record, as they may be updated and are annotated to the user as being derived. An astute reader will notice that simple chemical name lookup is not one of these methods; this is in part due to the aforementioned caveats.

The method using a PubChem identifier as a synonym is the most reliable, as the depositor already performed the curation step of matching their record to one existing within PubChem. The MeSH names are a controlled vocabulary used for name matching where expert analysts match references to chemicals in the biomedical literature to particular entries in the MeSH database. Provided that the link between CID and MeSH records is accurate, and if the depositor provided the correct synonym, this method is likely to provide good results for common names of substances. Name-to-structure software, although straightforward to use, has its own caveats. IUPAC (or IUPAC-ish) names have different styles and may contain some ambiguity, such that even different software packages or software versions may give different names for the same structure and different structures for the same name. Factor in the possibility of error when such names are generated by hand and IUPAC names begin to seem less reliable then some assume.

Incomplete Structure

The Protein Data Bank (PDB) is a classic case of a biological database resource with incomplete detail in small-molecule structure. Most PDB structures do not contain hydrogen, bond orders, or formal charges, as these are simply not part of the format. These details need to be inferred from the 3D geometry, using bond lengths, angles, and torsions to arrive at the atomic hybridization. This is accomplished automatically in PubChem with the help of the OEChem toolkit.[59] It is not perfect, however, because of ambiguities in the data such as 1UA0's "AF" ligand (SID 26711741), which comes out of OEChem (version 1.5.1) with a pentavalent carbon or when attempting to distinguish between the $NADP^+$/NADPH or FAD/FADH2 redox pairs, which have planar ring systems differing only in charge and the number of hydrogens, both of which are unspecified in the PDB format.

Literature extraction

To get chemical information from biological or other existing sources that have not maintained an explicit associated database of chemicals, it is often necessary to extract this information after the fact. Recognizing and cross-linking chemical references in biological journals and patent literature is an important and active field of development. The usual dichotomy exists between automated and manual extraction of this information meaning, is the text being processed by a computer or a human? Even determining which words in the text refer to chemicals is not a trivial problem, let alone ensuring that the right chemical structure is matched in each case. Especially with patents, both the results and the technology used may be proprietary. Published materials may include chemical structure drawings, which somehow need to be entered into a computer database. Technology commonly called chemical OCR is being developed on a number of fronts[60,61] to enable a computer to convert drawings automatically into detailed chemical structure representations.

9 CASE STUDY AND LESSONS LEARNED

We have presented a variety of technical challenges to building a chemical database and have used PubChem as an example of how to approach some of these problems. But in any database there will be errors, and one key to using a database properly is discerning whether errors originate in the data itself or in the database's particular infrastructure and algorithms. Let us keep these points in mind when considering the following example: how one might attempt to extract information on structure and biological activity from PubChem, based only on a chemical name.

When reading a journal article where detailed chemical structure is not present, we come across the name "vinblastine" and would like to discover what is known about this chemical. There are really two basic questions to explore here: What is the "correct" chemical structure of vinblastine, and what do we know about its biological activity? Are there relevant bioassay results in PubChem?

A simple unrestricted text search in the PubChem Compound database for this word will result in a list of 25 different compounds (at the time of this writing), all varying slightly in stereochemistry, salt form, or even basic formula and connectivity. As mentioned before, PubChem attempts to prioritize the search results so that structure most likely to be correct comes first in the search result. This is not a perfect algorithm, however, and is subject to the overall accuracy of the records from the numerous depositors (data sources) who have provided PubChem with its information. One might compare to, say, ChemSpider, in which a search for "vinblastine" results in a list of five structures. Careful comparison of these structures shows that the first structure from PubChem (CID 13342) exactly matches the first structure from ChemSpider (ID 12773), but ChemSpider does not seem to claim that the first record is the correct structure—it may just be coincidence.

Looking more closely at the variety of PubChem results, one sees in particular the structure of vindesine, also called (by MeSH) a "vinblastine derivative" or analog. Vindesine records are thus found by an unrestricted search for vinblastine, as PubChem makes MeSH descriptions part of the search. This is why it is important when doing a very specific search to narrow the search to appropriate fields. Using PubChem's CompleteSynonym index, only records where the search term exactly matches the (entire) name in a record will be found. So, searching PubChem Compound for "Vinblastine[CompleteSynonym]" results in 5 records, and excludes structures such as vindesine and vinblastine sulfate. Still, it seems unlikely that there are that many variants of this structure, as natural products such as this one tend to occur in only a single stereochemical form. These searches alone do not narrow the list enough to conclude which is the correct structure.

Putting aside (for the moment) the question of structure identity, let us turn to biological activity. As depicted in Figure 1, the compound at the top of the list in the initial search above (CID 13342) has a variety of information linked to it underneath the "Drug and Chemical Information" heading, including MeSH classification, DailyMed drug information, and safety and toxicology links. However, this particular record is not linked to any bioassays. One could go back to the exact name search and examine those records. But this might miss structures tested in assays but for which,

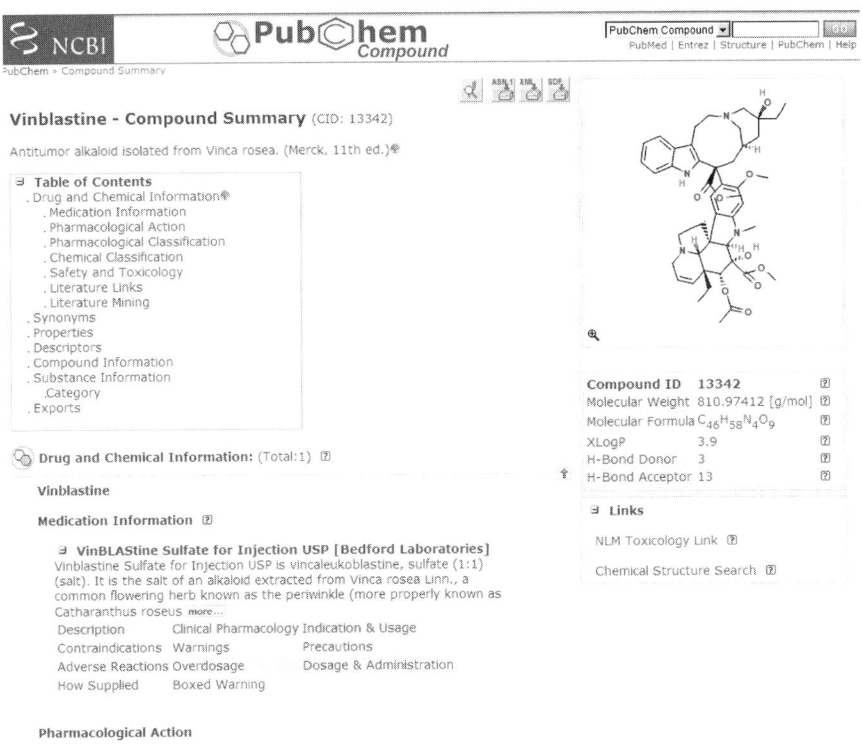

FIGURE 1 Partial view of the PubChem summary page for CID 13342, the nonsalt form of vinblastine.

for whatever reason, the name "vinblastine" was not given. To find such structures, one may use the PubChem links to compounds of the same chemical connectivity, found under the "Compound Information" heading on the summary page. From CID 13342, this link leads to a list of 19 compounds, 15 of which are called vinblastine, and two of these are tested in assays. However, since other ionic forms of the same structure are likely to have the same biological activity, one might better use the "same parent" links to expand the search; from an Entrez summary list from an earlier search, following the "Same Parent, Connectivity" link from the "Related Structures" pop-up menu for CID 13342 (see Figure 2), one arrives at a list of 46 compounds, seven of which have bioassay results. Indeed, at least one of these compounds with assay information (CID 16757894) does not have the name "vinblastine" associated with it at all.

From this Entrez result of 46 structures, let us use the bioactivity analysis tool to examine the bioassay results in more detail. Say that we are interested in compounds that are active in some assays. Activating the BioActivity Analysis button (containing two hexagons) near the top left

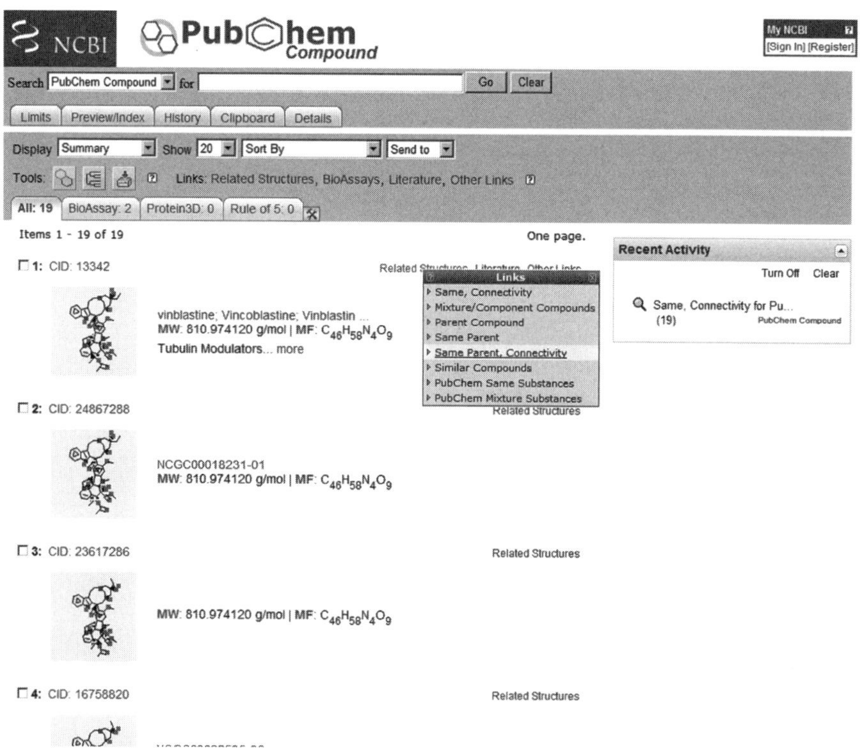

FIGURE 2 Partial view of an Entrez PubChem Compound display showing the 19 results having the same connectivity as CID 13342.

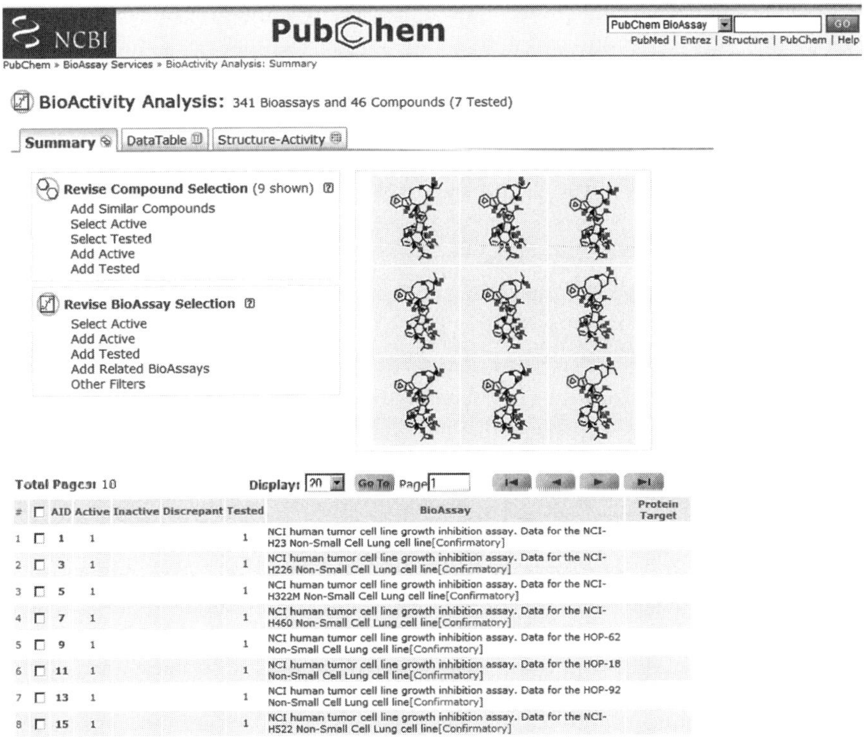

FIGURE 3 Partial view of a PubChem bioactivity analysis display for the 46 compounds that have the same parent compound connectivity to CID 13342. (*See insert for color representation of the figure.*)

of the Entrez result page, we get the analysis summary "341 Bioassays and 46 Compounds (7 Tested)" (see Figure 3), which is consistent with our previous Entrez search. In the default "Summary" tab of this tool, select the active compounds in the box labeled "Revise Compound Selection" and select the active assays in the box labeled "Revise BioAssay Selection." This will narrow the results to four compounds which were tested and found to be active in a total of 107 assays.

The entire list of all 107 assays may be shown by selecting "All" in the "Display" menu near the top. Note that in two assays, AIDs 589 and 590, there are both active and inactive results. In AID 589, CID 5388983 was found to be active, and this is the sulfate form of the original structure (CID 13342) found by the name search. CID 6604041 was found to be inactive and is different from CID 13342 at several stereocenters. It seems reasonable to hypothesize that the difference in activity is due to the difference in stereochemistry, assuming of course that the stereochemistry of these structures was represented correctly in the data supplied.

Going back to the list of 107 active assays, the "Structure Activity" tab of the analysis tool shows a graphical representation of the compounds clustered by structure. In this case the distribution of assay tests is mainly between two structures: CID 241902, tested in 91 assays supplied by NCI, and CID 5388983, in 13 assays from various Molecular Libraries centers. These two compounds (both sulfate forms) differ by a single bridgehead stereocenter, bringing us back to the question of the correct structure of vinblastine. Is this just a mistake in structure representation?

Resolving this question would probably require referring to original chemical literature on the isolation and structure elucidation of this natural product, which is beyond the scope of this work. But it is clear from this example that searching for information in a large database, especially one composed of data from multiple independent sources, is a nontrivial task. This is nothing new to the experienced chemical informatician. An effective search may require some slight fuzziness in textual or chemical structure search parameters to overcome data errors or differences in convention. PubChem has such a large number of "related structure" links for exactly this reason, so that a search may be tailored according to individual needs to make it as precise as necessary but at the same time flexible enough to enable discovery of information that is inexactly related, yet still relevant.

REFERENCES AND WEBSITES

1. http://pubchem.ncbi.nlm.nih.gov.
2. http://www.chembiogrid.org/related/resources/about.html.
3. *Chimia* **1998**, 52, 652 ff. Thanks to Dr. Kunz of the current *Chimia* editorial team for making this old issue available to us.
4. http://www.rcsb.org/pdb.
5. http://chembiofinderbeta.cambridgesoft.com; Brecher, J. S. *Chimia* **1998**, 52, 658.
6. http://webbook.nist.gov/chemistry.
7. O'Neil, M. J.; Heckelman, P. E.; Koch, C. B.; Roman, K. J., Eds. *The Merck Index: An Encyclopedia of Chemicals, Drugs, and Biologicals*, 14th ed., Merck & Co., Whitehouse Station, NJ, 2006.
8. Weinstein, J. N.; Myers, T. G.; O'Connor, P. M.; Friend, S. H.; et al. *Science* **1997**, 275, 343.
9. http://cactvs.nci.nih.gov/ncidb2; Ihlenfeldt, W. D.; Voigt, J. H.; Bienfait, B.; Nicklaus, M. C. *J. Chem. Inf. Comput. Sci*. **2002**, 42, 46.
10. Ertl, P.; Jacob, O. *Theochem* **1997**, 113.
11. http://www.jmol.org; Willighagen, E.; Howard, M. *CDK News* **2005**, 2, 17.

12. Sayle, R.; Milner-White, E. J. *Trends Biochem. Sci*. **1995**, 20, 374.
13. http://mdlchime.com/downloads.
14. http://www.biocheminfo.org/klotho; Kazic, T. *Biosystems* **1999**, 52, 111.
15. http://www.genome.jp/kegg; Ogata, H.; Goto, S.; Fujibuchim, W.; Kanehisa, M. *Biosystems* **1998**, 47, 119.
16. http://www.molecular-networks.com/biopath; Reitz, M.; Sacher, O.; Tarkhov, A.; Trümbach, D.; Gasteiger, J. *Org. Biomol. Chem*. **2004**, 2, 3226.
17. http://www.hmdb.ca; Wishart, D. S.; et al. *Nucleic Acids Res*. **2007**, 35, 521.
18. http://www.brenda-enzymes.info; Schomburg, I.; Chang, A.; Schomburg, D. *Nucleic Acids Res*. **2002**, 30, 47.
19. http://chem.sis.nlm.nih.gov/chemidplus; Wexler, P. *Toxicology* **2004**, 198, 161.
20. http://toxnet.nlm.nih.gov/cgi-bin/sis/htmlgen?TOXLINE Wexler, P. *Toxicology* **2004**, 186, 161.
21. http://toxnet.nlm.nih.gov/cgi-bin/sis/htmlgen?GENETOX; Wexler, P. *Toxicology* **2004**, 186, 161.
22. http://toxnet.nlm.nih.gov/cgi-bin/sis/htmlgen?CCRIS; Wexler, P. *Toxicology* **2004**, 186, 161.
23. http://chembank.broad.harvard.edu/welcome.htm; Seiler, K. P.; George, G. A.; Happ, M. P.; Bodycombe, N. E.; Carrinski, H. A.; et al. *Nucleic Acids Res*. **2008**, 36, 351.
24. Lipinski, C. A.; Lombardo, F.; Dominy, B. W.; Feeney, P. J. *Adv. Drug. Del. Rev*. **1997**, 23, 3.
25. http://pubchem.ncbi.nlm.nih.gov/pug/pughelp.html.
26. http://www.ebi.ac.uk/chebi/index.jsp.
27. http://www.drugbank.ca; Wishard, D. S.; Knox, C.; Guo, A. C.; et al. *Nucleic Acids Res*. **2006**, 34.
28. http://www.bindingdb.org/bind/index.jsp; Liu, T.; Lin, Y.; Wen, X.; et al. *Nucleic Acids Res*. **2007**, 35.
29. http://sw16.im.med.umich.edu/databases/pdbbind/index.jsp; Wang, R.; Fang, X.; Lu, Y.; Wang, S. *J. Med. Chem*. **2004**, 47, 12.
30. Zhang, J.; Aizawa, M.; Amari, S.; Iwasawa, Y.; et al. *Comp. Biol. Chem*. **2004**, 28, 401.
31. http://www.bindingmoad.org/; Hu, L.; Benson, M. L.; Smith, R. D.; Lerner, M. G.; Carlson, H. A. *Proteins* **2005**, 60, 333.
32. http://www.inpharmatica.co.uk/StARLITe.
33. http://nihroadmap.nih.gov.
34. http://www.ncbi.nlm.nih.gov/pubmed.
35. http://www.ncbi.nlm.nih.gov/Structure/MMDB/mmdb.shtml; Wang, Y.; Addess, K. J.; Geer, L.; Madej, T.; Marchler-Bauer, A.; Zimmerman, D.; Bryant, S. H. *Nucleic Acids Res*. **2000**, 28, 243.

36. http://www.ncbi.nlm.nih.gov/Genbank; Benson, D. A.; Boguski, M. S.; Lipman, D. J.; Ostell, J.; Oulette, B. F.; Rapp, B. A. M; Wheeler, D. L. *Nucleic Acids Res*. **1999**, 27, 12.
37. http://www.ncbi.nlm.nih.gov/sites/entrez?db=mesh; Humphrey, S. M. *J. Am. Soc. Inf. Sci*. **1984**, 35, 34.
38. http://dailymed.nlm.nih.gov/dailymed/about.cfm.
39. http://pubchem.ncbi.nlm.nih.gov/deposit.
40. http://www.chemspider.com.
41. http://www.mdli.com/downloads/public/ctfile/ctfile.jsp.
42. Weininger, D. *J. Chem. Inf. Comput. Sci*. **1988**, 28, 31–36; http://www.daylight.com/dayhtml/doc/theory/theory.smiles.html.
43. http://www.iupac.org/inchi.
44. http://pubchem.ncbi.nlm.nih.gov/sources.
45. Altschul, S. F.; Gish, W.; Miller, W.; Myers, E. W.; Lipman, D. J. *J. Mol. Biol*. **1990**, 215, 403.
46. http://blast.ncbi.nlm.nih.gov/Blast.cgi.
47. http://biocyc.org.
48. http://www.epa.gov/ncct/dsstox/index.html.
49. http://dtp.nci.nih.gov.
50. http://chemdb.niaid.nih.gov/struct_search/default.asp.
51. http://www.ncbi.nlm.nih.gov/books/bookres.fcgi/micad/home.html.
52. http://www.nature.com/nchembio.
53. http://nmrshiftdb.ice.mpg.de.
54. http://pubchem.ncbi.nlm.nih.gov/standardize.
55. Ihlenfeldt, W. D.; Gasteiger, J. *J. Comput. Chem*. **1994**, 15, 793.
56. http://pubchem.ncbi.nlm.nih.gov/search.
57. http://www.cas.org/expertise/cascontent/registry/regsys.html.
58. http://www.eyesopen.com/products/toolkits/lexichem-tk_ogham-tk.html.
59. http://www.eyesopen.com/products/toolkits/oechem.html.
60. http://cactus.nci.nih.gov/osra.
61. *J. Chem. Inf. Comput. Sci*., **1992**, 32(4), 373–378.

CHAPTER 2

BRIDGING CHEMICAL AND BIOLOGICAL DATA: IMPLICATIONS FOR PHARMACEUTICAL DRUG DISCOVERY

JEREMY L. JENKINS, JOSEF SCHEIBER, DMITRI MIKHAILOV, and ANDREAS BENDER
Center for Proteomic Chemistry, Novartis Institutes for BioMedical Research, Cambridge, Massachusetts

ANSGAR SCHUFFENHAUER
Center for Proteomic Chemistry, Novartis Institutes for BioMedical Research, Novartis Pharma AG, Basel, Switzerland

BEN CORNETT
Center for Proteomic Chemistry, Novartis Institutes for BioMedical Research, Cambridge, Massachusetts

VIVIEN CHAN
Oncology and Global Discovery Chemistry, Novartis Institutes for BioMedical Research, Emeryville, California

JASON KONDRACKI
Global Discovery Chemistry, Novartis Institutes for BioMedical Research, Emeryville, California

BERNHARD ROHDE
Center for Proteomic Chemistry, Novartis Institutes for BioMedical Research, Novartis Pharma AG, Basel, Switzerland

JOHN W. DAVIES
Center for Proteomic Chemistry, Novartis Institutes for BioMedical Research, Cambridge, Massachusetts

Computational Approaches in Cheminformatics and Bioinformatics, First Edition.
Edited by Rajarshi Guha and Andreas Bender.

1 INTRODUCTION

"Drowing in data and thirsty for knowledge"[1] is an often-quoted catch phrase in the information management field. Few industries can, however, claim this notion to be more pertinent to their current state than can the pharmaceutical industry. While the "omics" revolution (e.g., genomics, proteomics, metabolomics) is most frequently cited for its massive data output, other types of storage-intensive data, such as imaging data from high-content screening, are rapidly filling terabytes of disk space on pharmaceutical industry servers. Bioassay data using all flavors of perturbators—chemical, siRNA, shRNA, miRNA, and antibodies—as well as disparate assay technology types are collected and stored with the expectation of building a collective institutional knowledge base in each company. It is further expected that these data will be interoperable in many types of analyses. A number of looming questions are then pertinent: How do data become information and augment collective knowledge? How do big pharma companies—typically a patchwork of companies sewn together by a legacy of mergers and takeovers—ensure that they are working with suitably integrated data (e.g., where one protein name means the same thing throughout the company)? To what extent are the data generated by such expensive investments [$750,000 per high-throughput screening (HTS) campaign[2]] reused outside the original project team that collected the data? Are microarray data useful to a chemist? Are compound structure–activity relationships important to a bioinformatician? Finally, what are the strategies—and more important, the payoffs—for undertaking data integration initiatives? These questions are addressed throughout Chapter 3, in which we describe the current thoughts in the industry for bridging available chemical and biological data, with particular emphasis on lead finding. We follow a discussion on knowledge management with a review of currently available commercial chemical and biological databases and the efforts to integrate them with internal data in the pharmaceutical industry. A description of a Novartis effort to integrate diverse kinase data is highlighted in case studies as a successful example of using an environment where data silos are curated, standardized, and integrated where possible and otherwise federated by rule-based links. Finally, inspired by the comedian Steven Wright's sentiment, "experience is something you don't get until just after you need it," we follow with a "lessons learned" from our data integration initiative.

2 A CAUTIONARY TALE IN DATA PROVENANCE

Perhaps the single greatest leap in biology occurred in the 1830s when Charles Darwin—fueled by his observations as a naturalist on his expedition to the Galápagos Islands—proposed the theory of modified descent by natural selection. Darwin had collected a number of bird specimens from geographically isolated islands. As the story goes, he collected birds on each of the islands, noting that the bills of the birds had various shapes, yet were fashioned exquisitely for foraging techniques specific to the environmental conditions of each bird's island. Darwin—more of a geologist than a zoologist or bird anatomist—assumed that he had collected a number of different bird species. Once back in England, his error was corrected in 1837 by ornithologist John Gould, who identified the 13 or 14 birds as peculiarly distinct varieties of finch. Darwin then began to suspect that the finches could have evolved the bills most suitable to their environments by natural selection over time. A major glitch in his research, however, was that he failed to record the islands that his individual bird species hailed from.[2,3] Without this information, he could not support the hypothesis that the differing bills were adapted to particular environments. From an informatician's perspective, Darwin's predicament exemplifies the consequence of failing to capture data provenance or the original source of data (in this case, finch geography). Fortunately for Darwin (and the rest of us), crewmates on the HMS *Beagle* had a sporting interest in hunting birds on their trip, and they had recorded the locations of their avian trophies. After Darwin integrated his crewmates' data with his own observations of the finch bills, he then had the necessary knowledge to form his emergent theory of evolution by natural selection. In other words, his information became knowledge when contextualized by data provenance.

3 FROM DATA TO KNOWLEDGE

Although the ultimate business objective of the pharmaceutical industry is to produce and sell medicines, drugs as end products would be unobtainable without the vast knowledge generated en route to their discovery and development. Drug research and development (R&D) are both largely knowledge-producing exercises, and therefore the actual value of pharmaceutical industry is the knowledge generated. Acquiring institutional or even personal drug discovery knowledge from the bits and bytes captured by lab information management systems (LIMS) is nontrivial. First, it is critical to realize that data are not knowledge; it is not even information per se. Translation of data into knowledge is the goal of the emerging field of knowledge management (KM). KM is a special nuance of information technology (IT) which is rarely well appointed in pharma IT departments, and thus it is worth exploring these distinctions.

"Once data are organized they can be considered to be information. To derive knowledge from information, context and meaning must be attributed."[4] Put another way, "a lack of structure, not the amount, is the reason for our growing inability to cope with information today ... information has lost the connection with its carrier ...."[1,4] For scientific lab experiments, it is not uncommon for data organization to be carried out routinely by the principal scientists in their lab notebooks or in Microsoft Excel spreadsheets. The column titles in an Excel sheet are examples of information from organized data. Data from a high-throughput screen of 1 million compounds becomes information when the percent inhibition values for each compound are aligned with their corporate compound identification and chemical structure. Further, the biological target or cell line and the assay type, conditions, and technology may be annotated. When these data are scrutinized further by thoughtful scientists, knowledge may be gained by layering this information into the preexisting knowledge in their brains in order to give the information context. KM, however, contends that a similar degree of enlightenment must be obtained by a computer if the experimental results are to become stored, corporate knowledge. Context is then critical.

Machine learning is achievable only if semantically standardized terms are used to describe data and capture the full context of the experiment. For example, protein nomenclature may be taken from international standardizing bodies such as NCBI or Uni-Prot/Swiss-Prot (EMBL/EBI) rather than allowing free-typed protein names during assay registration. Whether enforced or enabled by software applications, semantic standardization of collected data enables mining to find patterns across disparate experiments or even the application of "inference engines." But importantly, semantically standardized data can only be created if scientifically meaningful controlled vocabularies, taxonomies, or ontologies (defined in the next section) are available to describe a scientific domain. In many cases they are not and must be custom-developed for niche scientific domains.

4 SEMANTIC INTEGRATION AND THE NEED FOR CURATION

Disparate data sources are an outgrowth of the rapidly changing landscape of scientific technologies. Despite the misleading term *drug pipeline*, research cannot be done linearly as on an assembly line; successful R&D relies on numerous disciplines being able to integrate their findings, as the separate drug pipeline issues cannot be resolved in isolation.[5] This means that the same "language" should be spoken in experiments spanning from biological target validation in early research, discovery, and optimization of chemical modulators, and assessment of safety and efficacy in animals and ultimately in humans. "Today's challenge is to resolve differences in meaning, or semantics, between disparate data sources."[6] The pharmaceutical industry "cannot expect to derive maximum benefit from the mere integration of its data without corresponding semantics."[7] In other words, standardization of data fields to which data results are attributed is essential to retain context and create interpretability beyond the scientist who performed the experiment.

As an example of a typical stumbling block to data integration, a recent analysis of drug side-effect annotation in the commercially available World Drug Index found no fewer than 15 versions of the term *hypersensitivity*, including multiple variations in capitalization and punctuation (e.g., "hypersensitivity," "Hypersensitivity," "HYPERSENSITIVITY," "hyper-sensitivity," "Hypersensitivity"), which may read the same to a human being but not to a machine learning algorithm. In total, 462 adverse drug reaction terms could be merged to 166 meaningfully distinct concepts.[8] Such lexical inconsistencies, even within a single data source, can make human learning laborious and machine learning impossible. The task of integrating multiple sources of such data becomes insurmountable without an army of curators.

There are a number of approaches to preventing or correcting inconsistent semantics, such as controlled vocabularies, taxonomies, and ontologies. A controlled vocabulary is a predefined lists of distinct terms that define the options for a data field. For example, the "assay type" for an HTS could be predefined as either "biochemical" or "cellular." A taxonomy is a hierarchical list of predefined terms with parent–child relationships. For example, a protein target taxonomy for the drug industry might begin with the parents "Enzyme," "Ion channel," "Nuclear Hormone Receptor," "GPCR," or "Other," where we would find children such as hydrolases and oxidoreductases for enzymes.[9] A number of target class–specific taxonomies have been developed based on either protein alignments or functional similarity.[10,11] In contrast, an ontology is a representation of all the concepts within a domain and the formal relationships between them.[1] Ontologies differ from taxonomies in that they need not be tree-shaped; they may have interconnected nodes more akin to a network or web. Ontologies make it possible to integrate silos of data by mapping equivalent entities. Large public (academic and government) initiatives have been launched in an effort to standardize life science ontologies.[12] The National Center for Biomedical Ontology is a consortium funded by the NIH roadmap to "develop innovative technology and methods that allow scientists to record, manage, and disseminate biomedical information and knowledge in machine-processable form."[13] NCBO has set up a repository called Open Biomedical Ontology (OBO, http://www.obofoundry.org/),[12] in which experts in the biomedical community may submit domain-specific ontologies. As of October 2008, 111 ontologies have been deposited, ranging from basic vertebrate anatomy to biological imaging methods to spider ontology. More relevant to the pharmaceutical industry is the Ontology for Biomedical Investigations (OBI), described by its consortium members as "an integrated ontology for the description of life-science and clinical investigations. This includes a set of "universal" terms, that are applicable across various biological and technological domains, and domain-specific terms relevant only to a given domain. This ontology will support the consistent annotation of biomedical investigations, regardless of the particular field of study. The ontology will represent the design of an investigation, the protocols and instrumentation used, the material used, the data generated and the type analysis performed on it" (http://obi-ontology.org/page/Main_Page).

Unfortunately, with some exceptions[14] it is rare to encounter data sources in which widely used ontologies were incorporated at their inception. This means that curation—or a combination of text mining and manual curation—is required retrospectively to "clean up" nonstandard data sources prior to integration. Given a large number of data silos typical in Big Pharma or even biotech, the level of resources required to integrate source data at the beginning of an integration project should not be underestimated.[15]

To address the issue of exactly what data should be captured from an experiment, the Minimum Information for Biological and Biomedical Investigations (MIBBI) initiative is seeking to incorporate OBO standards and to develop a process for consolidating the metadata necessary for data analysis.[16] If you imagine for a moment trying to list what information is really needed from an experiment to analyze the results, and repeating this for all possible assay types and technologies, you can begin to appreciate the enormity of the task of defining minimal information standards. Therein lies the rationale for an open-source public effort. Nevertheless, imposition of minimal information standards enables more powerful queries to databases that are not consigned to pulling back results, but serve to facilitate metadata creation from summary data. For example, the query "retrieve all compounds with valid IC_{50} results across all assays that used fluorescence detection" may offer you insights into the nature of chemical structures that are "frequent hitters" in such assays in order to flag them as potential false positives. This information goes outside the rationale for conducting each individual chemical screen and presents nascent knowledge that would be difficult to obtain without a controlled vocabulary or taxonomy for assay technologies. (It is worth noting that taxonomies for HTS technologies are still needed in the public domain.)

5 KNOWLEDGE CURATORS

Recently, Howe et al. observed that "biocuration, the activity of organizing, representing, and making biological information accessible to both humans and computers, has become an essential part of biological discovery and biomedical research."[17] A biocurator handles tasks such as reading full-text articles, extracting data, resolving synonyms, tagging with controlled vocabularies, fixing inconsistencies, and loading data into a database. There is even a growing trend toward biocuration as a career path. There is yet to be a similar push for "chemocuration," with the notable exceptions of ChEBI,[14] ChemSpider,[18,19] and Wikipedia, and to a lesser extent NCBI PubChem and DrugBank.[20] For example, to the authors' knowledge, there is no Ph.D.-track program for cheminformatics in the United States. In contrast, as of 2008, there are at the very least 36 American universities offering Ph.D. programs in bioinformatics, computational biology, or medical informatics.[21] This has to do in part with the large public initiative to sequence the human genome and the subsequent explosion of academic bioinformaticians trained to analyze large-scale genomic data. Equivalent analyses of large-scale chemical data have been consigned to the pharmaceutical industry where large corporate compound collections are available. As the largest chemical curator, Chemical Abstracts Services, has not pursued a business model that allows the release of their chemical structures *en masse*, it is only in recent years that online chemical databases or even commercially licensed chemical databases have been available for cheminformatics research.[22] Even so, controlled vocabularies implemented in early chemical databases such as MDDR (MDL Elsevier) and the World Drug Index (Thomson) to annotate compound targets were out of sync with parallel bioinformatics efforts.

In practice, the constant output of new methodologies and life science domains is outpacing the availability of informaticians that fully understand both the science as well as the IT issues required for knowledge management. In our experience, bio- and cheminformaticians—arguably the bio- and chemocurators of the industry—are scarce in their placement in disease areas and functional areas of pharmaceutical companies.

Further, it is not uncommon for informaticians to be focused on software infrastructure in IT development groups rather than being embedded in project teams, where they can most directly affect drug discovery.

6 COMMERCIAL CHEMICAL DATABASES

Databases of biologically annotated chemical structures have long been essential for cheminformatics research, especially for method development in the areas of structure–activity relationship (SAR) and quantitative SAR analyses, machine learning, probabilistic modeling, and high-throughput docking. In effect, chemical structures annotated with biological activities form a chemogenomics knowledge base. A *knowledge base* is "a special kind of database for knowledge management" that "provides the means for the computerized collection, organization, and retrieval of knowledge."[23] A chemogenomics knowledge base consists of chemical structures, such as SMILES (Simplified Molecular Input Line Entry System)[24] or InChIs (IUPAC International Chemical Identifier), annotated with established biological activities. Activities are typically quantitative and range from straightforward biochemical target SAR values (IC_{50}, K_i, EC_{50}, K_D, etc.) to more complex and less-defined cellular or animal phenotype results. Annotations for chemical databases are most often hand-curated from medchem literature and patents, especially for commercial databases. At present, such hand-curated ligand–target pairings are on the order of $1.00 per record and available from more than half a dozen vendors (Table 1).

Ideally, target annotations conform to controlled vocabularies.[25] Even more useful are target annotations based on standard nomenclature from organizing bodies, which facilitate semantic integration with other data sources. Whether or not controlled vocabularies are used determines to what extent automated analyses may be applied to a knowledge base. For example, a human may guess that ABL1, Abl kinase, and c-Abl are probably the same concept, whereas a computer may not distinguish among them without a specific algorithm to resolve synonyms. Commerical databases may also be distinguished by their access methods. In particular, certain Web-access-only licensing models may limit data-mining approaches relative to full-access download approaches (Figure 1).

Historically, all cheminformaticians were heavily reliant on a few data sources; in particular, the MDL Drug Database Report dominated as a control set for model training and docking in methodology publications from the mid-1990s to the early 2000s. The influence of data set composition bias by recycling a single data source was understood only later.[26] In the last few years, however, both public and commercial efforts have yielded

TABLE 1 Commercial Chemical Databases with Broad Target SAR Data[a]

Database	URL	Description	Size	Data Source Coverage	Breadth of Target Coverage	Target Names Standardized for Data	Target Ontology	GUI
GVK target inhibitor (patents)	http://www.gvkbio.com/informatics.html	Large, diverse collection of chemical series from medchem literature and patents	1.8 Mio entries, 500K compound records, 1.5K targets	Journals and patents	Wide (enzymes, ion channels, GPCRs, nuclear hormone receptors, others)	Yes, official gene name of protein	Yes	GOSTAR Web access, ISIS Base, Oracle, MS-Access
GVK MedChem (literature)	http://www.gvkbio.com/informatics.html	Large, diverse collection of chemical series from medchem literature	607K unique compounds, 4.9K targets	64 medicinal chemistry journals	Wide (enzymes, ion channels, GPCRs, nuclear hormone receptors, others)	Yes, official gene name of protein	Yes	GOSTAR Web access, ISIS Base, Oracle, MS-Access
Aureus AurSCOPE GPS	http://www.aureus-pharma.com/Pages/Products/Aurscope.php	Collection of "knowledge bases": GPCR, kinase, ion channel, protease, nuclear receptor	500K compounds, >3K targets	~370 journals and thousands of patents	Focused on major target classes (general "Enzymes" not included yet)	Yes	Yes	AurQuest (Web based)

(continued)

TABLE 1 ***(Continued)***

Database	URL	Description	Size	Data Source Coverage	Breadth of Target Coverage	Target Names Standardized for Data	Target Ontology	GUI
EMBL-EBI stARLITe (purchased from Galapagos in 2008)	None available yet; check: http://www.ebi.ac.uk/	Large, diverse collection of chemical series from medchem literature	300K compounds, ~5K targets, 1.3 million datapoints	Medchem journals (1980–present)	Wide (enzymes, ion channels, GPCRs, nuclear hormone receptors, others)	Yes, RefSeq or ACCESSION and some standardized gene names	Yes	Unknown at this time
Prous Science Integrity	http://www.prous.com/integrity/	Compounds under investigation in patents, conference proceedings, biomedical literature, medical trials	278K compounds, 2050 targets	Normally, Web access only unless full-dump licensed	Wide (enzymes, ion channels, GPCRs, nuclear hormone receptors, others)	Yes, UniProt IDs	Yes	Integrity Web interface
Sunset Molecular WoMBat	http://www.sunsetmolecular.com/index.php/wombat	Chemical series published in medchem literature	192.5K unique compounds and 1979 protein targets	4 journals (10,205 papers)	Wide (enzymes, ion channels, GPCRs, nuclear hormone receptors, others)	Yes, normalized names and ACCESSION	Yes	ISIS Base or SDF

GeneGo MetaDrug	http://www.genego.com/metadrug.php	Pharmacological data curated from literature	~500K compounds? Thousands of targets	64 medicinal chemistry journals	Wide (enzymes, ion channels, GPCRs, nuclear hormone receptors, others)	Yes	Yes	MetaCore or MetaDrug
Jubilant BioSys Chem-BioBase	http://www.jubilantbiosys.com/chembio.html	Combined target-centric ligand databases: Kinase, GPCR, Nuclear Receptor (NR), Ion Channel, Protease, Phosphodi-esterase	~ 1020 total targets. kinase, 319K compounds; GPCR 400K compounds; NR 150K compounds; ion channel, 100K compounds, protease, 400K compounds	Standard medicinal chemistry journals and patents	Focused on large target classes (general "Enzymes" not included)	Yes	Yes	Isis Base
Cerep BioPrint	http://www.cerep.fr/cerep/users/pages/ProductsServices/bioprintservices.asp	Focused pharma-cology/ADME profiling database; a full data matrix	Full matrix of 2,400 compounds and 158 diverse targets	Experimentally determined in vitro data	Wide, but small number	Yes	Not needed	BioPrint GUI

[a]Table information is extracted from the URL websites in September 2008 or product literature.

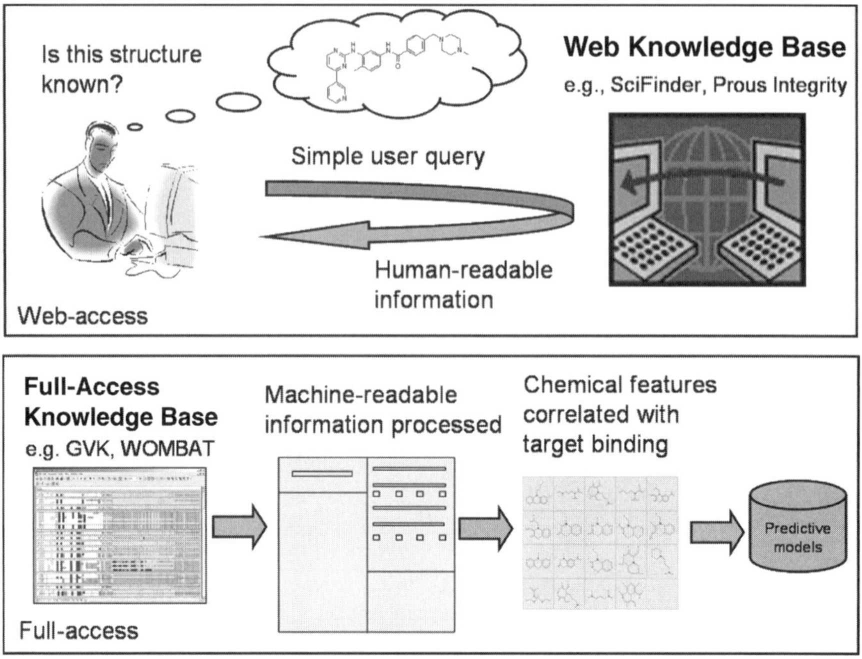

FIGURE 1 Web-access knowledge bases (top) are designed for specific user queries, typically with limited downloads or exports of query results. Their main advantage is potentially helpful visualizations, whereas data mining approaches are not possible. Full-access knowledge bases (bottom)—if structured with machine readable annotations—enable the training of statistical models that create additional knowledge and can be used for predictions.

databases that cover a substantial portion of the ligand–protein interactions from patents and publications. Public efforts in chemical database curation, such as PubChem, are covered in Chapter 1. Despite growing efforts in the academic and government sectors to create large-scale chemical knowledge bases, there remains almost no effort outside commercial vendors to curate ligand–target relationships from patents, arguably a more difficult task than curating medchem papers or deposited HTS data. The greater difficulty is due to the nature of how patents are written. Patent writing is a conflicted task of maximizing chemical space coverage for legal intellectual property protection while obfuscating chemical structures as much as possible to make elucidation a more frustrating task for competitors. Chemotypes are typically represented with Markush structures (a core scaffold with R-groups), where enumerating all possible combinations presents a combinatorial explosion of structures that make for a disagreeable chore of curation, particularly if the activity is not for profit. Efforts to parse compound–target pairings from patents with automated text-mining approaches have not yet achieved the same level of accuracy as manual curation, a task not perceived as cost-effective by most pharmaceutical companies. Thus, there is a market for commercial chemical databases based on hand-curated patents.

The largest browser-oriented commercial chemical bioactivity database is the CAS database, searchable by SciFinder. The CAS Registry contains over 38 million compounds. In an effort to make their data more minable, CAS have developed a scaffold-based approach to analyzing exported sets of records from SciFinder called SubScape. SubScape scaffolds are based on the Murcko and Bemis graph frameworks[27] and can be visualized by biological targets, activities, and companies. DiscoveryGate (Symyx) is a Web-access enterprise platform for searching chemical structures from 20,000 journal titles plus patents. Integrity (Prous Science) is a Web-access portal to well-integrated drug R&D information covering chemistry, genomics, tissue engineering, pharmacodynamics, pharmacokinetics, and pathology. Custom taxonomies for nearly every category were developed by Prous and are exposed in the GUI to help the user conduct queries.

Full-access commercial chemogenomics databases are listed in Table 1, the largest in terms of compounds and targets is the GVK database suite. Where chemical structures or SMILES are linked to standardized target names, statistical models may be trained broadly to find correlations between chemical features and target activity. These mappings of pharmacological space enable a novel way of comparing proteins—by their chemical space similarity[28–30]—and allow prospective predictions of targets for chemical structures,[31,32] annotation of chemical libraries,[33,34] and assessment of IP coverage *en masse*.

7 BIOLOGICAL DATABASES FOR FEDERATION AND ANNOTATION

The vast number of bioinformatics resources available online makes it nearly impossible to list them. Indeed, the 2007 Nucleic Acids Research Database issue exceeded 1000 databases. Most important to understand is which databases are fully downloadable and which ones contain widely known standard identifiers to which other databases can be federated or joined by rules-based links. Table 2 lists a number of key resources for annotating protein targets, many of which are public and downloadable in their entirety. The possible annotations from these sources range from structure and function information for proteins to expression and localization information, and to pathways, phenotypes, and disease relationships.

TABLE 2 Biological Databases Useful for Annotating Targets

Database	Source	Availability	Description	Key Concept for Annotating Small-Molecule Targets
EntrezGene	NCBI	Public	"nomenclature, map location, gene products and their attributes, markers, phenotypes, and links to citations, sequences, variation details, maps, expression, homologs, protein domains and external databases" [Maglott et al, 2005]	Entrez Gene ID and Gene Symbol, Gene Description
OMIM (Online Mendelian Inheritance in Men)	NCBI	Public	compendium of human genes and genetic phenotypes in Mendelian disorders	OMIM ID, Disease
UniProtKB	UniProt Consortium (EBI, SIB, PIR)	Public	Protein sequence and function from Swiss-Prot (manually curated) and TrEMBL (automatically curated)	UniProt/SwissProt ID; Accession Number, protein
Interpro	EMBL-EBI	Public	Protein families, domains, repeats, sites	IPR ID
HPRD (Human Protein Reference Database)	Johns Hopkins University	Public with industry restrictions	". . . a centralized platform to visually depict and integrate information pertaining to domain architecture, post-translational modifications, interaction networks and disease association for each protein in the human proteome" [Mishra et al. 2006]	Entrez Gene ID, SwissProt ID, OMIM ID, RefSeq ID, GenBank ID
BRENDA	T. U., Braunschweig, Germany	Public through Web, commercial full access	Hierarchical classification of enzymes by function	EC numbers
MEROPS	Wellcome Trust Sanger Institute	Publicly licensed	Hierarchical classification of peptidases, includes substrates and protein inhibitors,	MEROPS identifiers

GO (Gene Ontology)	GO Consortium, >21K members or associates	Public	gene processes, function, location using a controlled vocabulary [Nature Genetics 2000]	GO ID and terms
KEGG (Kyoto Encyclopedia of Genes and Genomes) PATHWAY	Kanehisa Laboratories (Kyoto, Tokyo)	Public with industry restrictions	Maps and modules for molecular interaction and reaction networks	KEGG Pathway ID
Reactome	Cold Spring Harbor, EBI, GO, Rutgers, UCSD	Public	Curated knowledge base of pathways and protein–protein interactions	Reactome Stable Identifier
RSCB/PDB		Public	. structures from x-ray, NMR, and electron microscopy	PDB Codes
Gene Expression Atlas	Genomics Institute of the Novartis Foundation	Public for noncommercial use	Tissue expression of genes	Tissue and abundance
Mouse Genome Informatics (MGI)	Jackson Laboratories	Public	Mutations, phenotypes, orthology	Phenotype Summary
MetaCore, MetaDrug, MetaBase (database)	GeneGo	Commercial (with full-access option)	Curated biological pathways	GeneGo Pathways
IPA	Ingenuity	Commercial (GUI access)	Curated biological pathways	Not available outside GUI
SRS	BioWisdom	Commercial (GUI or programmatic access)	Gene and protein sequences, various annotations—full bioinformatics platform	Various

Commercial biological databases that can be found quite often in corporate settings are GeneGo's MetaCore and Integrity's IPA for pathway analysis and BioWisdom's SRS for sequence retrieval and annotation. Examples of integrating biological and chemical databases will follow in a subsequent section describing a Novartis kinase data integration project.

8 BIG PHARMA CHEMICAL AND BIOLOGY INTEGRATION EFFORTS

Reported attempts to integrate chemical and biological data in large pharmaceutical companies reflect either the use of tools to amalgamate data or the construction of integrated databases with varying degrees of sophistication. The most often cited integration tool for supporting such work is data pipelining software: namely, Pipeline Pilot (Accelrys, SciTegic) and KDE (InforSense). Pipeline Pilot was released in 2000 and is used by such companies as Amgen, Sanofi-Aventis, Dupont, Johnson & Johnson, Merck, Millenium, Novartis, and Pfizer. InforSense has cited collaborations with AstraZeneca and GlaxoSmithKline to break data silos and work across all of R&D to push translational research.[35] Data pipelining tools are described in greater detail in Chapter 9. Pipelining tools are particularly important for annotating large compound library files. Database integration efforts in pharma span from local knowledge bases to companywide enterprise systems. Integrating chemogenomics databases from multiple vendors or public sources is itself a task, independent of additional biological data. AstraZeneca published an analysis of the overlap of PubChem, GVKBio, BioPrint, DrugBank, MDDR, WOMBAT, and others[36] using the OpenEye toolkit. Zhou et al. at the Genomics Institute of the Novartis Research Foundation (GNF) created an "annotation integration pipeline" workflow for annotating compound collections based on structure matches to various data sources containing bioactivity information. This approach sheds light on hits in bioassays and also enabled a MeSH ontology-driven analysis of GNF compounds with respect to their global HTS profiles.[34]

In their seminal report on "global mapping of pharmacological space,"[30] Hopkins and colleagues at Pfizer cited integrating biologically active chemical structures from both internal and external screening data, as well as commercial databases from stARLITe (Inpharmatica), BioPrint (Cerep), and IDDB (Thomson). Importantly, assays were mapped to targets by protein sequence, and indications were annotated with a disease taxonomy from the Medical Dictionary for Regulatory Activities (MedDRA). A number of emergent analyses resulted from this integration, such as a dendrogram of target relationships in chemical space and profiling of compound properties by target family. In reference to integrating disparate biological information, Huang and colleagues, also at Pfizer, have taken the contrary stand that "data integration is not the answer."[7] Instead, they advocate knowledge systems that incorporate a scheme for representing facts for information domains in a relational database linked to knowledge acquisition systems, which are further enhanced with algorithms for searching and visualization and an expert reasoning system for hypothesis generation.

Eli Lilly has also used integrated biological and chemical data to support chemogenomics approaches, particularly with respect to kinases ("kinomics").[37,38] They have explored the relationship of kinase SAR to kinase identity and linked kinases by chemical substructure similarity. On a different subject, Eli Lilly has been leading noticeably in efforts to integrate the semantic web into life sciences, the semantic web being another form of integration not discussed in this chapter but likely to be critical to future discussions.[39]

Johnson & Johnson have discussed their Advanced Biological and Chemical Discovery (ABCD) informatics platform, consisting of a data warehouse, an application suite for querying, mining, and reporting, and a workspace for data sharing and collaboration.[40] The warehouse encompasses pharmacological bioassay data, HTS and secondary data, sample management, project tracking, and more. Importantly, the warehouse makes generous use of dimensional modeling to introduce data redundancy, enabling faster and simpler queries. They note that the most difficult aspect of data warehousing is determining the rules for populating the integrated database. To assist this process they employed commercial tools for ETL (extraction, translation, and loading) and ontologies from HUGO, GO, ATCC cell line repository, and MeSH, but were forced at times to develop their own classification schemes. Of course, manual intervention was still needed at times ("all existing targets were manually curated").

In a 1990s effort to increase organizational memory, Millennium researchers, together with Ingenuity, described their MyBiology initiative, designed to resolve the problem of IP blind spots and to integrate internal and external scientific findings, with the ultimate goal of encouraging users to think about mapping from compound to pathway to disease rather than the traditional compound-to-target paradigm.[41] It is an example of early semantic integration approaches to life science data.

Vertex has developed many in-house capabilities, such as the Vertex In-house Research Database (VERDI). It is a knowledge management application containing historical information about compounds, enabling a scientist working on a current compound structure to receive alerts for historical activity values, potential liabilities, and so on.[42]

Novartis has used its text-mining expertise to integrate a vast array of full-access databases, ranging from genes, proteins, and assays, to drug and company information and more, in a tool called UltraLink. As described in Bio-IT World in September 2006: "The contents of each UltraLink result page are read and categorized by an expert system at loading time. This enables the selection of pertinent pages based on a treelike representation of extracted concepts and entities. UltraLinks are created that associate each extracted entity with a set of meaningful links to other databases and applications. ... Key entities—genes, companies and institutions, diseases and indications, etc.—are color-coded by category and highlighted on returned pages." While UltraLink linked information successfully by ontologies, another initiative at Novartis, the Novartis Data Federation Initiative (NDFI), sought to integrate chemical and biological data as a data retrieval system for complex scientific queries.

9 THE NDFI KINASE PILOT

Everyday scientific questions asked by researchers often cross interdisciplinary boundaries with a boundlessness not amenable to data silos. Without comprehensive oversight, the lack of standardization, annotation, and accessibility of data makes it impossible to ask challenging but important scientific questions. In the same spirit of the declaration put forth by Hopkins et al. that the "correct compilation and integration of chemical and biological data is the foundation that transforms drug discovery into a knowledge-based predictive science,"[30] we set forth to create an integrated chemical and biological data environment from both internal and external sources. Specifically, the exercise was limited to kinase data only in the proof-of-concept stage, as kinase biology and chemistry are rich sources of disparate data.

Having a vision for database content requires the database designer to anticipate the types of questions that may be asked of the data. Rather than leaving this up to guesswork or imposing our own personal biases, we collected a list of questions from researchers and key opinion leaders throughout the company. The questions helped to ensure that the project stayed focused on real-life problems, as they were pertinent to real work on kinases and could not be answered easily by querying data silos. Moreover, the questions were used to inform the architecture of a new data environment containing multiple data sources and concept dictionaries and taxonomies. The objective was to assess the extent to which integrating data could enable efficient and timely answers to the questions. Following is a representative sample of the questions.

1. What are the inhibitors of kinase X and the five most similar kinases with $IC_{50} < 1\ \mu M$ and with $MW < 500$ from all internal and external data sources?
2. What assay technologies have been used against my kinase? Which cell lines?
3. What other proteins are in the same kinase branch as target X, where there were validated chemical hits from external or internal sources?
4. If I hit a particular kinase, what would the potential side-effect profile look like? Which known inhibitor of this kinase has the best safety profile and the fewest known IC_{50}s?
5. Have I identified other compounds with a bioactivity profile similar to compound X and with the same core substructure?
6. Can we create a phylochemical tree of kinases and for a new kinase target place it into the tree on the basis of activity against a reference panel of compounds?
7. Have I identified all kinases with an x-ray structure (in-house or external) that are in pathway X?

10 NDFI: FEDERATION VS. WAREHOUSING

Data integration tactics come in many flavors, from simple linking of data fields between data silos to full-scale amalgamation into a single warehouse environment. Simply put, data federation allows the querying of data across multiple sources without relocation of the original data. In contrast, data integration requires porting data to one location via ETL scripts. Federation requires some commonalities between sources in order to join them. Commonalities need to be resolved either at the primary source or by "middleware frameworks" that resolve differences.[15]

For our kinase data integration project, we opted for a combination of warehousing and federation. This was a reasonable strategy for a number of reasons.

1. Chemical bioassay or high-throughput screening data can be reasonably standardized and integrated into a warehouse, as the nature of the experiment is generally the same in all cases: A compound at some concentration is used to perturb a target, cell, or phenotype using an assay with a quantitative readout that can be normalized to a percent inhibition value

or IC_{50}-type values from fitted dose–response curves. In other words, there are reusable concepts for storing data that apply *across* multiple assays. Bioassay data from a number of sources—internal, outsourced, public, commercially licensed—was available and not previously integrated. Additionally, annotation of assays to standard identifiers for protein name was lacking. These facts compelled us to create a data warehouse for kinase small-molecule data from multiple internal and external bioassay databases in a single standardized format and to employ manual curation where needed. For example, target Entrez Gene Symbols were curated for all assays, as were assay type (cell or biochemical) and assay technology. The latter required creation of a new assay technology taxonomy, as nothing similar was available in the public domain; this is not unexpected due to the relatively recent advent of HTS in academic settings.

2. Many bioinformatics resources were available that were not kinase specific but were nevertheless well annotated or conformed to public taxonomies. For example, we generally found that either NCBI/Entrez GENESYMBOL and GENEID or Uni-Prot/Swiss-Prot Accession Numbers were used in nearly every biological database to represent proteins (or another identifier that could be mapped to them). Although small-molecule screening databases lacked such standard protein identifiers, biological databases were naive to chemical structure representations: If present, IUPAC chemical names, drug names, or CAS numbers were used rather than canonical SMILES or InChIs, the common currencies of cheminformaticians for chemical structure description. That said, it was nearly trivial to decide to federate from the small-molecule warehouse to these proteomics, genomics, disease, and structural databases rather than to try to integrate them into a single environment.

3. As pointed out in a 2005 review article on data integration[43]: "Combining databases, by whatever means, requires recognizing entities (such as genes or compounds) that are common to domains. This task can be straightforward if there are simple unique keys by which to recognize the equivalence of entities, and preferably standardized vocabularies or even universal identifiers...." By picking a standard target identifier and standard chemical identifier, data federation enabled a quick win, given that no new application would be needed to interface to the data sources. For targets, we selected Entrez GENE ID, and for compounds we selected SMILES canonicalized by an in-house algorithm. CAS numbers were not used, as we wished to include proprietary chemical matter. Thus, the small-molecule bioassay database was federated to many existing databases (e.g., proteomic, structural) scattered around Novartis using standardized target names and canonical SMILES (Figure 2). We also

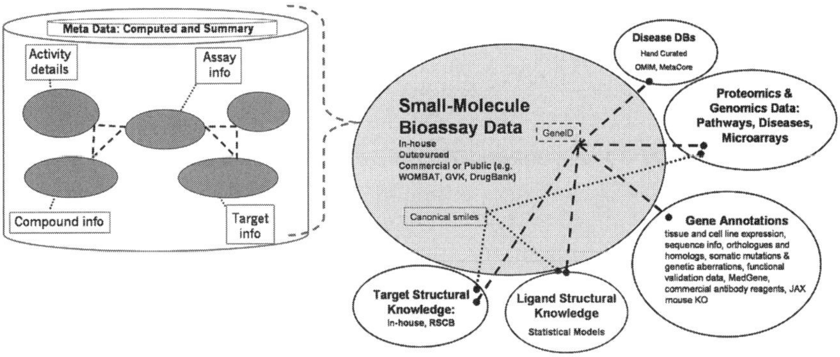

FIGURE 2 Mixed model of data integration and data federation used in the Novartis kinase data integration project. Creation of a small-molecule bioassay knowledge base (gray circle) of in-house, licensed, public, outsourced, and computed results required a nascent, semantically-organized relational data warehouse, organized as in the cylinder at left. The dotted-line triangles shown between concept nodes indicate critical relationships that need to be captured: projects-assays-targets and assays-compounds-results. Assays form the common intersection of these. The projects-assays-targets relationship captures the *context* of assays, whereas assays-compounds-results relationships captures *what happened* in the assay. Given that each concept can be further annotated (e.g., targets linked to target ontologies or cell lines linked to cell line dictionaries), the intersection of these two triangles makes the difference between a knowledge base and a database. At right, the small-molecule knowledge base is federated to other pre-existing data sources via canonical SMILES for chemical structures and Entrez GENE ID for human kinases.

converted kinase-specific disease databases, an in-house x-ray structure database, and other public data sets to the same standardized formats.

4. Pipeline Pilot (Accelrys, SciTegic) could easily be set up to access all data sources that we wished to federate, using either ODBC connections or flat-file text readers. Importantly, the powerful data pipelining environment of Pipeline Pilot allowed us to forgo developing an application or GUI (graphical user interface) with "metasearch" capabilities, which would require extra resources and could have derailed the time line of a proof-of-concept study.

Case Study 1: From Cellular Pathways to Small-Molecule Inhibitor Conformations

The intersection of chemistry and biology data often breaks down where compound activities for targets could in principle easily be related to targets and molecular pathways. One query collected in the NDFI list was "Get all kinases with an x-ray structure (in-house or external) that are in pathway *X*," where *X* is user defined. Such a question may arise when looking for compound antagonists in a phenotypic or pathway screen. Triaging hit lists from pathway screens can be difficult, due to the ultimate task of identifying a target for compound hits. Having the information above could help in two ways. First, having co-crystal structures of proteins with ligands from your pathway of interest may enable annotation of screening hits that resemble—either in two- or three-dimensional chemical space—known inhibitors. For example, by similarity searching or pharmacophore searching the pathway screening hits with co-complexed ligand structures as reference compounds, one might elucidate the molecular targets. Second, one could dock the pathway screen hits to the x-ray structures of targets in the pathway as further support for a target hypothesis. Alternatively, one could forgo a pathway screen and attempt to use structural information available for pathway targets for lead finding.

Figure 3 shows the workflow used in the NDFI to answer the pathway-to-structure question. It's important to remember that no workflow existed prior to the NDFI to facilitate answering the question. Using data pipelining, four different source databases were accessed in a single workflow: (1) a pathways database (PAT) where pathways are linked to gene IDs and pathways come from either GeneGo, KEGG, or others; (2) the NDFI database to distinguish kinases from other targets; (3) RSCB, external PDB structures; and (4) in-house crystal structures.

Case Study 2: Compound–Target Cladistics: Growing a Kinome Phylochemical Tree

Cladistics is a field of study in evolutionary biology that seeks to determine hierarchical classification of species, often using quantitative genomic sequencing data. Ancestral lineages of organisms can be visualized by placing them on dendrograms (referred to specifically in this context as cladograms or phylogenetic trees). The original *kinome tree* published by Manning et al. at Sugen in 2002[10] is based on the similarity of catalytic domain sequences. It has been used widely since its inception to frame conversations about kinases based on the eight branches of the tree. The tree is also used both explicitly and implicitly in lead-finding approaches, where inhibitors of kinases genetically related to the kinase of interest are used as starting points or tool compounds.

In drug discovery, the chemogenomics approach takes advantage of evolutionary biology knowledge for lead finding.[44] While the molecular similarity principle states that "like compounds have like activities,"[45] the principle can also be extended to the idea that "like targets bind like compounds."[46,47] Thus, there is a reciprocal relationship that can be used by drug researchers when only minimal information is known about a

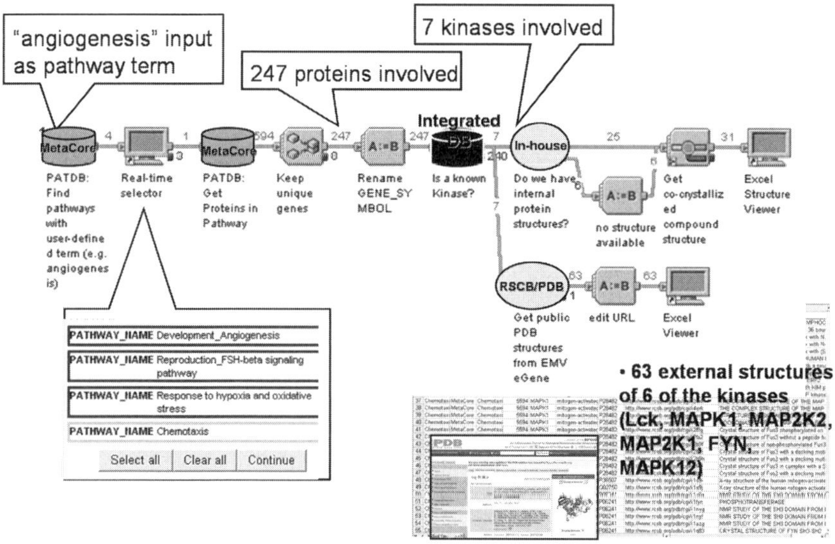

FIGURE 3 From pathway to target to compounds co-crystallized with kinases: an example of using data pipelining to connect federated databases. Pipeline Pilot (Accelrys) ODBC connections were created for all federated databases. First, all pathways containing the protein EGFR are retrieved and narrowed down to the MetaCore (GeneGo) pathway "chemotaxis". Subsequently, all proteins in the chemotaxis pathway are retrieved and filtered to include only kinases. Kinases with x-ray crystal structures, complexed with compounds in both in-house or external databases, are retrieved and exported as batch lists. The protocol took roughly 20 mins to construct and seconds to run, whereas reconstructing the same information from accessing each data silo individually would take days of work.

new drug target; looking at the chemical inhibitors of genetically similar protein targets may jump-start lead finding as long as the new target's sequence is available for BLASTing a database of ligand–drug targets. In this sense, the chemogenomics approach is "underwritten" by evolution: Nature has painstakingly shaped proteins for specific functions, and where these functions and shapes are similar to one another, phylogenetic *by proxy* SAR can be exploited for lead finding. Drug targets that are similar genetically have structurally similar binding sites, which is the basis for compound cross-reactivity. Sometimes, however, the phylogenetic tree is not entirely predictive of how compounds will behave. That is, compounds will not always inhibit neighboring nodes on the cladogram tree. This seems to be more true for kinases than for GPCRs.[46] Or, to the converse, distant nodes on the tree may be inhibited by similar chemotypes. When genetically dissimilar proteins bind to similar ligands, one might consider it a *homoplasy*, a character shared by multiple tree members by some cause other than common ancestry.[48] Typically, homoplasies occur due to convergent evolution. Examples of homoplasies can be found in chemical-based drug target trees, such as the similarity of inhibitors of dissimilar enzymes that use folate.[29] Given, the extensive compound–target data that are available for kinases and the only modest correlation in kinase similarity to kinase SAR, one query collected in the NDFI questions list was: "Can we create a new tree for kinases based on chemical SAR and place a new kinase target on this tree?"

Using the NDFI small-molecule kinase database in Figure 3, we dumped the collective set of kinase-compound $IC_{50} < 5$ μM values where IC_{50}, and at least 10 compounds were found per kinase. The set was used to train a ligand-based multiple-category naive Bayesian model (Figure 4). Pipeline Pilot ECFPs (extended connectivity fingerprints) were used as the chemical descriptors. In this manner, chemical features and substructures that are associated statistically with activity to each kinase were determined along with a statistical weight for each feature. The chemical features and weights for each kinase model were then compared to one another by Pearson correlation to yield a kinase

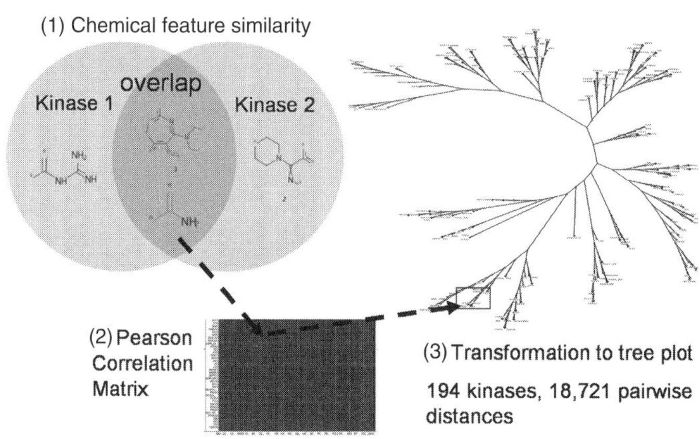

FIGURE 4 Construction of a kinome phylochemical tree based on similarity of chemical features. Multicategory Bayesian models were trained using Pipeline Pilot ECPFs (extended connectivity fingerprints) chemical descriptors for each kinase where >10 compound structures with IC_{50} values <5 μM are known. The chemical features/substructures statistically associated with activity to each kinase are determined along with a statistical weight for each feature. The Bayesian weights place on each chemical feature associated with kinase activity were compared in a full Pearson similarity matrix. A normalized similarity was computed (equal to $-0.5 * \text{Pearson} +0.5$). Similarities between each kinase were then transformed to a tree view using the KITSCH method (Phylip v3.6.7) and displayed in PhyloDraw v0.82 as a radial tree.

similarity matrix, which is then converted to similarity values:

$$\text{normalized similarity} = -0.5 \times \text{Pearson} + 0.5$$

The similarity matrix was subsequently transformed into a tree representation using the KITSCH method (Phylip v3.6.7) and displayed in PhyloDraw v0.82 as a radial tree (Figure 4). How does one apply the phylochemical tree versus the phylogenetic tree for lead finding? In other words, how can we use these trees to predict hits for a new kinase? We applied the knowledge in the phylochemical tree to an active Novartis project for a kinase target with only three known inhibitor chemotypes—all established in-house. There was no knowledge from the literature or patents available for the kinase with respect to chemical matter. The three chemotypes in-hand were profiled by small-molecule affinity chromatography[49] to find both on- and off-target kinases pulled down from cellular lysate. Similarly, the chemotypes were profiled in a panel of kinase biochemical assays to look for off-target activities. Interestingly, these off-target data—normally perceived as helpful for understanding safety and selectivity—could be exploited to place the kinase on the phylochemical kinome tree, which enabled us to pick compounds that are known or predicted to inhibit the multiple "off-target" kinases hit by the three chemotypes. In effect, the off-targets serve as a collective inverse pharmacophore for the desired on-target. Since not all compounds in our library have been tested against all kinases, we chose to build a tree neighbor Bayesian model (Figure 5) based on our phylochemical similarities. For comparison, we also created a tree neighbor Bayesian model based on kinome phylogenetic similarities. The workflow for creating a tree neighbor model is simple. First compounds that are known or predicted to inhibit neighbors on the tree are pulled from the database. The compounds are used in aggregate as the "active set" to train a ligand-based Bayesian model. The model then represents the chemical space associated statistically with the entire branch of interest. For example, for the phylogenetic tree, one could have a chemical feature-based TK branch Bayesian model to represent all tyrosine kinases or a TKL branch Bayesian model for all tyrosine kinase–like kinases. In practice, phylochemical tree neighbors of a given kinase are the off-target kinases found by profiling of the chemotypes known to hit the kinase of interest. These are supplied, as mentioned previously, by small-molecule affinity chromatography or biochemical profiling in kinase panels.

How do the two trees perform for prospective lead finding? We tested several thousand compounds selected from the corporate library by both

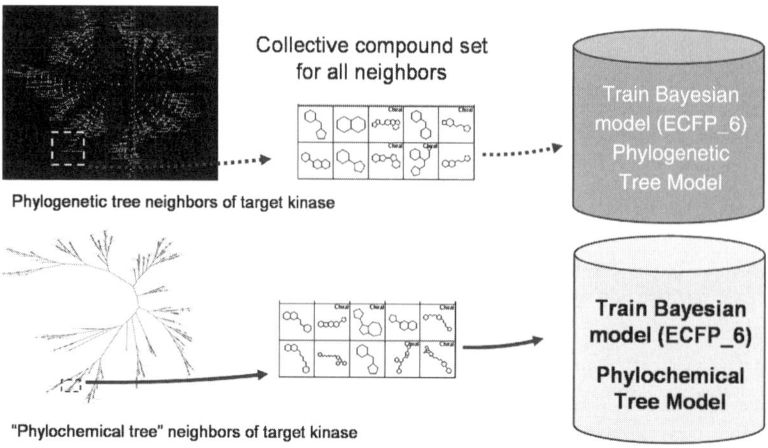

FIGURE 5 Workflow for chemogenomics-driven lead finding. Based on phylogeny of a new protein target to other known targets, inhibitor compounds can be found. In this example, the position of a kinase on a sequence-based phylogenetic tree is used to find its neighboring branch nodes. The compound inhibitors of the tree neighbors are aggregated and used to train a Bayesian classification model, which is then used to rank a chemical library for potential leads. Targets with at least 1 or 2 tool compounds may be placed on a phylochemical tree by profiling the tool compounds in kinase assay panels or by small-molecule affinity chromatography to determine common "off-targets". These off-targets enable placement on the kinase-of-interest on a phylochemical tree. The workflow then follows that of the previous approach. (*See insert for color representation of the figure.*)

the phylogenetic neighbor model and the phylochemical neighbor model. It is important to note that we removed all compounds from our selection with prior known kinase activity, in an effort to encourage chemotype novelty in the hit list. At the same time we screened a 10,000-compound diversity set. In our prospective experiment, the compounds from the diversity set yielded a hit rate of 0.8%. In contrast, compounds selected by the phylogenetic model yielded a fivefold better hit rate than the diversity hit rate. Based on a scaffold analysis of the hits, fewer than 50% of the hits had completely novel scaffolds. The phylochemical model resulted in 16-fold better hit rate than the diversity set and considerably more than 50% novel scaffolds. Thus, in our single example, the phylochemical tree was a more powerful predictor of new hits than the protein sequence–based tree.

The advantage of a phylochemical tree approach for lead finding is that the process is fully automatable, highly scalable, and fast once a kinase knowledge base is in place. The enriched hit rates enables successful small focused screens. The disadvantage is that a ligand–target knowledge base must be available for mining. Second, totally novel targets with no reference inhibitors cannot be plotted onto a phylochemical tree, whereas the phylogenetic model approach only requires a protein sequence.

All in all, it is critical to note that recycling prior "off-target" knowledge can drive new lead discovery and that target inhibition profiles constitute a biological signature that strongly predicts activity against a single target.

11 LESSONS LEARNED

A number of lessons were learned over the course of the NDFI. First, there are no assay ontologies in the public domain that fully describe pharmaceutical research, particularly for HTS. However, guidelines for screening and the nature of data collected have been issued by a joint effort of Eli Lilly and the National Institutes of Health.[50] Minimal information standards need to be developed specifically for high-throughput screening technologies to ensure proper data analysis. For example, it is critical for future data mining tasks that "assay rationale" be captured by the experimentalist. Rationales may encompass such concepts as primary screen, confirmation screen, validation, counterscreen (to test for assay artifact), and selectivity screen. Additionally, rules for what "IC_{50}" means and what criteria are necessary to judge the validity of a dose–response curve are important for large-scale filtering of "good" $IC50_s$s from "bad" ones.

Automating knowledge capture is a very different task from curation of historical documents and requires a more complex understanding of ETL (extraction, translation, and loading) and a set of rules for how to handle exceptions to rules. Regardless, curating backward in time is often necessary to bring historical experiments up to standard. Curating should only be carried out up to the current day; going forward, the deficient processes that created the need for curation must be fixed. This typically involves using controlled vocabularies, taxonomies, or ontologies maintained by a central, responsive group of curators. Data standardization does not need to be implemented at the level of LIMS systems but, rather, at the level of warehousing data sources. ETL scripts that are savvy to controlled vocabularies and help to translate source data to a useful format for warehousing are critical and must be supported by property-resourced IT scientists. As may be expected, there are many technical and bureaucratic barriers to integrating research data, and even worse, for integrating research and development or clinical data. It is important to create incentives for such tasks as assay annotation rather than to enforce compliance where possible. For example, "free text" options in assay capture systems should be minimized in favor of simple drop-down menus with controlled vocabularies.

Recent advocates of "peer production" projects such as Wikipedia exploit the power of the public and a growing desire to be "prosumers" rather than consumers of goods.[51] However, we find that this optimistic view extends only so far in a company setting, and that "ontology by committee" is probably not a good idea. In practice, a mixed model of small, agile core teams is better for initial formation of ontologies, with the option of getting feedback from disease area scientists to evolve the ontologies.

In our experience, IT projects with a "build it and they will come" mentality are generally doomed to failure in large pharmaceutical environments. This is particularly true of Web services such as cheminformatics predictions created by software development groups, particularly if they are not "advertised" widely. It is our opinion that informatics success comes from three components: (1) a good infrastructure reflecting the principles described in this chapter, (2) innovative science, and (3) direct application of informatics approaches and knowledge to active projects. Given a good infrastructure, the greatest impact an informatician can have is on the ground and in the trenches with experimentalists. There is no substitute for being embedded in project teams to acquire a real understanding of what is needed to drive the science forward. In practice, an applied cheminformatics group that provides project-specific expertise will have far greater impact than tools developed with the by "backroom" modelers that are released on an unknowing public.

12 SUMMARY

Scientific data must be organized to become useful information, which in turn becomes knowledge when enriched with context. For life science experiments, annotation of concepts by ontologies are providing a way to give experiments a universal context that is understandable by computers. Of course, "the best people to do annotation are the people who are producing this knowledge in the first place."[52] However, biocurators may retrospectively help to add context to enable data mining.

Integration of assay-compound-activity results together with assay-target-project metadata from disparate historical data sources can answer complex scientific questions. For example, questions that expand beyond individual experiments can be answered with the emergent knowledge gained from semantically integrated bioassay data: Which compounds are frequent hitters in fluorescence assays (assay technology)? In GPCR family assays (protein ontology)? In biochemical assays (assay types)?

What similar targets could be exploited for my focused screen (protein sequences in addition to SAR)?

Finally, integrated data can drive probabilistic lead finding using chemogenomics principles (e.g., using phylogenetic or phylochemical similarities). Importantly, the chemogenomics paradigm relies on "cross-assay" knowledge, which is only possible when assays are appropriately annotated, standardized, and accessible in full.

REFERENCES

1. Koniger, P.; Janowitz, K. Drowning in information, but thirsty for knowledge. *Int. J. Inf. Manage*. **1995**, 15, 5–16.
2. Bender, A.; Bojanic, D.; Davies, J. W.; Crisman, T. J.; Mikhailov, D.; Scheiber, J.; Jenkins, J. L.; Deng, Z.; Hill, W. A.; Popov, M.; Jacoby, E.; Glick, M. Which aspects of HTS are empirically correlated with downstream success? *Curr. Opin. Drug Discov. Dev*. **2008**, 11(3), 327–337.
3. Zimmer, C. *Evolution: The Triumph of an Idea*. Harper Perennial, New York, 2002.
4. Torr-Brown, S. Advances in knowledge management for pharmaceutical research and development. *Curr. Opin. Drug Discov. Dev*. **2005**, 8(3), 316–322.
5. Pisano, G. P. Can science be a business? Lessons from Biotech. *Harvard Bus. Rev*. **2006**, Oct.
6. Bergman, M. K. Climbing the Data Federation Pyramid. http://www.mkbergman.com [internet blog], **2006**, p. 229.
7. Slater, T.; Bouton, C.; Huang, E. S. Beyond data integration. *Drug Discov. Today* **2008**, 13(13–14), 584–589.
8. Bender, A.; Scheiber, J.; Glick, M.; Davies, J. W.; Azzaoui, K.; Hamon, J.; Urban, L.; Whitebread, S.; Jenkins, J. L. Analysis of pharmacology data and the prediction of adverse drug reactions and off-target effects from chemical structure. *ChemMedChem*. **2007**, 2(6), 861–873.
9. Schuffenhauer, A.; Zimmermann, J.; Stoop, R.; van der Vyver, J. J.; Lecchini, S.; Jacoby, E. An ontology for pharmaceutical ligands and its application for in silico screening and library design. *J. Chem. Inf. Comput. Sci*. **2002**, 42(4), 947–955.
10. Manning, G.; Whyte, D. B.; Martinez, R.; Hunter, T.; Sudarsanam, S. The protein kinase complement of the human genome. *Science* **2002**, 298(5600), 1912–1934.
11. Webb, E. C. *Enzyme Nomenclature: Recommendations of the Nomenclature Committee of the International Union of Biochemistry and Molecular Biology on the Nomenclature and Classification of Enzymes*. Academic Press, New York, 1992.

12. Smith, B.; Ashburner, M.; Rosse, C.; Bard, J.; Bug, W.; Ceusters, W.; Goldberg, L. J.; Eilbeck, K.; Ireland, A.; Mungall, C. J.; et al. The OBO Foundry: coordinated evolution of ontologies to support biomedical data integration. *Nat. Biotechnol*. **2007**, 25(11), 1251–1255.
13. Rubin, D. L.; Lewis, S. E.; Mungall, C. J.; Misra, S.; Westerfield, M.; Ashburner, M.; Sim, I.; Chute, C. G.; Solbrig, H.; Storey, M. A.; et al. National Center for Biomedical Ontology: advancing biomedicine through structured organization of scientific knowledge. *OMICS* **2006**, 10(2), 185–198.
14. Degtyarenko, K.; de Matos, P.; Ennis, M.; Hastings, J.; Zbinden, M.; McNaught, A.; Alcantara, R.; Darsow, M.; Guedj, M.; Ashburner, M. ChEBI: a database and ontology for chemical entities of biological interest. *Nucleic Acids Res*. **2008**, 36(Database issue), D344–D350.
15. Gardner, S. P. Ontologies and semantic data integration. *Drug Discov. Today* **2005**, 10(14), 1001–1007.
16. Taylor, C. F.; Field, D.; Sansone, S. A.; Aerts, J.; Apweiler, R.; Ashburner, M.; Ball, C. A.; Binz, P. A.; Bogue, M.; Booth, T.; et al. Promoting coherent minimum reporting guidelines for biological and biomedical investigations: the MIBBI project. *Nat. Biotechnol*. **2008**, 26(8), 889–896.
17. Howe, D.; Costanzo, M.; Fey, P.; Gojobori, T.; Hannick, L.; Hide, W.; Hill, D. P.; Kania, R.; Schaeffer, M.; St Pierre, S; Twigger, S; White, O; Rhee, S.Y. Big data: the future of biocuration. *Nature* **2008**, 455(7209), 47–50.
18. Williams, A. J. Public chemical compound databases. *Curr. Opin. Drug Discov. Dev*. **2008**, 11(3), 393–404.
19. Williams, A. J. Internet-based tools for communication and collaboration in chemistry. *Drug Discov. Today* **2008**, 13(11–12), 502–506.
20. Wishart, D. S.; Knox, C.; Guo, A. C.; Cheng, D.; Shrivastava, S.; Tzur, D.; Gautam, B.; Hassanali, M. DrugBank: a knowledge base for drugs, drug actions and drug targets. *Nucleic Acids Res*. **2008**, 36(Database issue), D901–D906.
21. http://www.cbi.or.jp/cbi/DigtalTL/Bioinfo_Universities.html. 2008.
22. Guha, R.; Schurer, S. C. Utilizing high throughput screening data for predictive toxicology models: protocols and application to MLSCN assays. *J. Comput. Aided Mol. Des*. **2008**, 22(6–7), 367–384.
23. http://en.wikipedia.org/wiki/Knowledgebase. 2008.
24. Weininger, D. SMILES, a chemical language and information system: 1. Introduction to methodology and encoding rules. *J. Chem. Inf. Comput. Sci*. **1988**, 28, 31–36.
25. Jenkins, J. L.; Bender, A.; Davies, J. W. In silico target fishing: predicting biological targets from chemical structure. *Drug Discov. Today: Technol*. **2008**, 3, 413–421.
26. Bender, A.; Mussa, H. Y.; Glen, R. C. Screening for dihydrofolate reductase inhibitors using MOLPRINT 2D, a fast fragment-based method employing the naive Bayesian classifier: limitations of the descriptor and the importance

of balanced chemistry in training and test sets. *J. Biomol. Screen*. **2005**, 10(7), 658–666.

27. Bemis, G. W.; Murcko, M. A. The properties of known drugs: 1. Molecular frameworks. *J. Med. Chem*. **1999**, 39, 2887–2893.
28. Keiser, M. J.; Roth, B. L.; Armbruster, B. N.; Ernsberger, P.; Irwin, J. J.; Shoichet, B. K. Relating protein pharmacology by ligand chemistry. *Nat. Biotechnol*. **2007**, 25(2), 197–206.
29. Hert, J.; Keiser, M. J.; Irwin, J. J.; Oprea, T. I.; Shoichet, B. K. Quantifying the relationships among drug classes. *J. Chem. Inf. Model*. **2008**, 48(4), 755–765.
30. Paolini, G. V.; Shapland, R. H.; van Hoorn, W. P.; Mason, J. S.; Hopkins, A. L. Global mapping of pharmacological space. *Nat. Biotechnol*. **2006**, 24(7), 805–815.
31. Nidhi; Glick, M.; Davies, J. W.; Jenkins, J. L. Prediction of biological targets for compounds using multiple-category Bayesian models trained on chemogenomics databases. *J. Chem. Inf. Model*. **2006**, 46(3), 1124–1133.
32. Young, D. W.; Bender, A.; Hoyt, J.; McWhinnie, E.; Chirn, G. W.; Tao, C. Y.; Tallarico, J. A.; Labow, M.; Jenkins, J. L.; Mitchison, T. J.; Feng, Y. Integrating high-content screening and ligand-target prediction to identify mechanism of action. *Nat. Chem. Biol*. **2008**, 4(1), 59–68.
33. Plouffe, D.; Brinker, A.; McNamara, C.; Henson, K.; Kato, N.; Kuhen, K.; Nagle, A.; Adrian, F.; Matzen, J. T.; Anderson, P.; et al. In silico activity profiling reveals the mechanism of action of antimalarials discovered in a high-throughput screen. *Proc. Natl. Acad. Sci. USA* **2008**, 105(26), 9059–9064.
34. Zhou, Y.; Zhou, B.; Chen, K.; Yan, S. F.; King, F. J.; Jiang, S.; Winzeler, E. A. Large-scale annotation of small-molecule libraries using public databases. *J. Chem. Inf. Model*. **2007**, 47(4), 1386–1394.
35. May, M. Working out the flow. *Bio-IT World*, September 18, 2006. Cambridge Healthtech Institute, Needham, MA USA.
36. Southan, C.; Varkonyi, P.; Muresan, S. Complementarity between public and commercial databases: new opportunities in medicinal chemistry informatics. *Curr. Top. Med. Chem*. **2007**, 7(15), 1502–1508.
37. Sutherland, J. J.; Higgs, R. E.; Watson, I.; Vieth, M. Chemical fragments as foundations for understanding target space and activity prediction. *J. Med. Chem*. **2008**, 51(9), 2689–2700.
38. Vieth, M.; Higgs, R. E.; Robertson, D. H.; Shapiro, M.; Gragg, E. A.; Hemmerle, H. Kinomics: structural biology and chemogenomics of kinase inhibitors and targets. *Biochim. Biophys. Acta* **2004**, 1697(1–2), 243–257.
39. Feigenbaum, L.; Herman, I.; Hongsermeier, T.; Neumann, E.; Stephens, S. The Semantic Web in action. *Sci. Am*. **2007**, 297(6), 64–71.
40. Agrafiotis, D. K.; Alex, S.; Dai, H.; Derkinderen, A.; Farnum, M.; Gates, P.; Izrailev, S.; Jaeger, E. P.; Konstant, P.; Leung, A.; et al. Advanced

biological and chemical discovery (ABCD): centralizing discovery knowledge in an inherently decentralized world. *J. Chem. Inf. Model*. **2007**, 47(6), 1999–2014.

41. Ficenec, D.; Osborne, M.; Pradines, J.; Richards, D.; Felciano, R.; Cho, R. J.; Chen, R. O.; Liefeld, T.; Owen, J.; Ruttenberg, A.; et al. Computational knowledge integration in biopharmaceutical research. *Brief. Bioinf*. **2003**, 4(3), 260–278.
42. Ertl, P.; Seltzer, P.; Muhlbacher, J. Web-based cheminformatics tools deployed via corporate intranets. *Drug Discov. Today: Biosilico* **2004**, 2(5), 201–207.
43. Searls, D. B. Data integration: challenges for drug discovery. *Nat. Rev. Drug Discov*. **2005**, 4(1), 45–58.
44. Jacoby, E.; Schuffenhauer, A.; Floersheim, P. Chemogenomics knowledge-based strategies in drug discovery. *Drug News Perspect*. **2003**, 16(2), 93–102.
45. Maggiora, G. M.; Johnson, M. A. *Concepts and Applications of Molecular Similarity*. Wiley, New York, 1990.
46. Kuhn, M.; Campillos, M.; Gonzalez, P.; Jensen, L. J.; Bork, P. Large-scale prediction of drug–target relationships. *FEBS Lett*. **2008**, 582(8), 1283–1290.
47. Schuffenhauer, A.; Floersheim, P.; Acklin, P.; Jacoby, E. Similarity metrics for ligands reflecting the similarity of the target proteins. *J. Chem. Inf. Comput. Sci*. **2003**, 43(2), 391–405.
48. West-Eberhard, M. *Developmental Plasticity and Evolution*. Oxford University Press, New York, 2006.
49. Bantscheff, M.; Eberhard, D.; Abraham, Y.; Bastuck, S.; Boesche, M.; Hobson, S.; Mathieson, T.; Perrin, J.; Raida, M.; Rau, C.; et al. Quantitative chemical proteomics reveals mechanisms of action of clinical ABL kinase inhibitors. *Nat. Biotechnol*. **2007**, 25(9), 1035–1044.
50. http://www.ncgc.nih.gov/guidance/manual_toc. html. 2008.
51. Tapscott, D.; Williams, A. D. *Wikinomics: How Mass Collaboration Changes Everything*. Portfolio, New York, 2006.
52. Waldrop, M. Big data: Wikiomics. *Nature* **2008**, 455(7209), 22–25.

CHAPTER 3

CHEMOINFORMATICS TAKING BIOLOGY INTO ACCOUNT: PROTEOCHEMOMETRICS

JARL E. S. WIKBERG, OLA SPJUTH, MARTIN EKLUND, and MARIS LAPINS
Department of Pharmaceutical Biosciences, Uppsala University, Uppsala, Sweden

1 INTRODUCTION

Biological and medicinal chemistry research is characterized by massive growth in the amount of experimental data becoming available. This is largely due to the advent of automated equipment for data acquisition: in biology the application of "omics," including genomics, transcriptomics, proteomics, metabonomics, and interactomics, and in medicinal chemistry the application of high-throughput screening of large collections of synthetic and natural compounds and automated methods in chemical library synthesis. All of this has led to an immense need to find computational methods that can take advantage of the growing amount of information, and of computerized methods that can organize and make the data retrievable in easy and user friendly ways. The area of research that deals with this is called *e-science* and constitutes "the systematic development of research methods that exploit advanced computational thinking" and pertains to all areas of science, not only biology and chemistry. Thus far, resources for e-science have been allocated largely to hardware, like clusters, whereas the resources to develop the theories, algorithms, and software that can make e-science the revolution that it promises to become are lagging

Computational Approaches in Cheminformatics and Bioinformatics, First Edition.
Edited by Rajarshi Guha and Andreas Bender.

behind. At present we are therefore still at the very beginning of the path that we have to follow to take full advantage of all the opportunities that are given by the massive amounts of data accumulating in modern biological and chemical research.

One of the areas in which the deficiencies in the level of current e-science become immediately evident is the area of drug discovery and pharmaceutical development. While the genomics revolution has brought about at least 3000 new drug targets, only about 300 of these have been utilized for drugs and less than five new chemical entities with a novel mechanism of action enter the market each year, with the time from lead to marketed drug averaging 14 years and the costs being proportionate.[1,2] At this rate it will take quite some time to fulfill all the possibilities, and the costs will be astronomical.

The task in drug discovery is to tailor a number of concomitant, often contradictory properties into a single drug molecule. This includes finding a molecule that has the proper activity on the target(s) selected, while not causing any unwanted "off-target" effects. The compound should be nontoxic, it should be possible to formulate into a pharmaceutical preparation that can be delivered for the treatment of a subject, and once it has entered the body it should reach the target in sufficient concentration to exert its pharmacological action. On top of that it should also show reasonably good pharmacokinetic properties, have a rate of metabolism and excretion that is not too fast or too slow, and should preferably be novel (i.e., not be covered by previous patents or otherwise reside in the public domain).

The problems of contemporary pharmaceutical development are due largely to the fact that it proceeds along a linear time axis where very many candidates are tried, and most of them sorted away during the development, until a final candidate is deemed to be suited for clinical use. Moreover, although prior knowledge from earlier developmental projects is placed into any new project, this is done in quite a random fashion, in a way that is largely dependent on the mind-set of the developers who happened to be available for the development project and on what was comprehended by them from the proprietary records accumulated from the earlier developments and the scientific data publicly available. Overall this is a suboptimal approach which has a low chance of success unless very many initial candidates are being entered into this drug developmental "funnel." A more optimal approach would be to feed *all* available prior knowledge into a project using algorithmic approaches that in an unbiased fashion guide the developer to make optimal choices at any point in the development, and then iteratively feeding back all project data accumulated, thus improving each step of the development in case they need to be retaken. Thus, in short this would be to use a systematic statistical learning approach in drug development.

Today, once a drug target has been selected, the drug discovery process usually begins with high-throughput screening in the range of 10^5 to 10^6 compounds to find "leads,". This is then followed by brute-force trial-and-error synthesis of large arrays of compounds to refine these leads. However, in silico approaches have increased the efficiency of high-throughput screening and include computational approaches for prefiltering of chemical libraries by docking of structures into the active site of the target, pharmacophore screening and other approaches. This can reduce the size of the screening library while retaining the number of hits. Once lead compounds are found, methods based on three-dimensional (3D) docking can be applied further to optimize the interactions of the leads with the target. Even in the absence of leads, de novo ligand design can be performed provided that an accurate 3D structure of the target is available. However, obtaining high-resolution structures for biomacromolecules using x-rays or NMR is often problematic, particularly so for integral proteins, which confines most of the desired drug targets. Moreover, 3D structures are merely static models of the proteins and can account only poorly for the conformational flexibility of the target and the structural changes caused by ligand binding and complexation with other proteins, and this is difficult to remedy even if one applies time and resources consuming molecular dynamic simulations to simulate such flexibility.

Another widely applied computational method in drug discovery is quantitative structure–activity relationship (QSAR), which is used to optimize leads for target activity and other properties (ADME, toxicity, off-target activities). However, since the major drug targets are members of large groups of related proteins (e.g., G-protein-coupled receptors and protein kinases), using single QSAR models may be insufficient to attain a drug with the desired specificity. Further, many diseases have very complex pathophysiology, and effective treatment requires regulation of multiple cellular signaling cascades, prompting the need for multiselective drugs. Yet another aspect that must be taken into account is the ability of many targets to undergo mutational changes (e.g., many pathogens respond to treatment by developing resistance mutations in the target or target-associated proteins) that render the pathogen insensitive to the treatment. Similar resistance patterns due to mutational changes also occur in cancer. Current technologies in drug design are poorly adopted to deal with such "moving" targets, where the target undergoes constant structural changes due to mutations.

In view of the discussion above it is thus obvious that current technologies for drug development need to be improved. Although no single measure alone can solve all these problems, we have for some time been developing a promising approach—proteochemometrics—which may correct many of the deficiencies of the current mainstream drug-design technologies. Proteochemometrics was conceived by us in 1999[3] and has since been developed and validated in a number of publications.[4–11] In proteochemometrics one analyzes the interaction data for many target proteins simultaneously, and proteochemometric models are created by correlating the chemical properties of proteins and their ligands (organic compounds, peptides, etc.) to the interaction data measured. This reveals properties that are important for protein–ligand complementarity, which can be used to design drugs with improved activities and selectivities. Proteochemometric models may take advantage of the fact that biomolecules that have diverged functionally from the same ancestral gene show sequence similarity and share essentially the same tertiary structure. This makes it possible to correlate chemical properties of primary amino acid sequences directly to protein–ligand interactions. Using proteochemometrics in such a fashion, the targets' 3D structures are not needed, and proteochemometrics may even be capable of accounting for long-range interactions and conformation flexibility of protein interactions. However, current trends in the development of proteochemometrics indicate that its scope may be even more far reaching than this. This is because unified proteochemometric models were recently made over broad groups of target proteins that do not show any obvious sequence or fold similarity.[8,12]

The aim of this chapter is to review the status of proteochemometrics and to introduce new software in development that can be used to effectively manage proteochemometrics data and modeling. Moreover, we discuss future directions that can be taken where proteochemometrics is integrated in a larger framework, taking into account prior biological data from heterogeneous data sources.

2 PROTEOCHEMOMETRICS

Definition and Principles

As the name suggests, proteochemometrics originates from chemometrics, a broad field of research that has been developed gradually since the 1970s, which deals with all types of mathematical and statistical aspects of chemistry. In proteochemometrics, chemometric approaches to, for example, design of experiments, description of molecules, statistical learning, and validation of models are extended from purely chemical data to biological data and systems. A definition of *proteochemometrics* was given in 2004 as the area where the molecular interactions between proteins and their interacting entities are studied using chemometric techniques.[5] To make such studies possible, both the proteins and the interacting entities are described by numbers that have bearing on their physical and chemical properties, and empirical relations are deduced from them that serve as a model for their interactions. In this way proteochemometrics becomes a generalization of quantitative structure–activity relationship (QSAR) (Box 1). While QSAR models the interactions of a group of molecules with only one target, proteochemometrics affords simultaneous modeling of the interactions of a group of molecules with several targets. As we discuss in detail below, this joint perturbation of both the ligand and the target yields additional information about interactions between proteins and ligands. The advantage of proteochemometrics is that it not only deals with the chemical space defined as the chemical property space spanned by all molecules of a series of ligands, as QSAR does, but also with the interaction space defined as the space spanned by the noncovalent interactions between all molecules of the targets and the interacting entities. As a result, proteochemometrics produces more descriptive and stable models with higher predictability than QSAR.[13] However, proteochemometrics is not necessarily limited to studying two interacting entities only. It could involve more interactors, such as protein subunits, protein–protein interaction, other types of biomacromolecules (e.g., DNA, RNA), environmental factors, and more. However, from a practical point of view it is limited to the availability of suitable data, and so far most studies cover proteins interacting with peptides or small organic molecules.

BOX 1: THE CONCEPT OF PROTEOCHEMOMETRICS

Proteochemometrics is an approach to the study of the interaction space of interacting molecular entities. Similar to QSAR, it applies descriptors to molecules that consist of numerical coding of molecular properties. Descriptors can be calculated from a molecule's structure or determined experimentally, and they can range from very simple (such as a compound's molecular weight, count of atoms, number of rotatable bonds), to very complex (such as being computed from the ability of probe atoms to form noncovalent interactions with the molecule in three dimensions).

Quantitative structure–activity relationships (QSARs) are limited to relating chemical descriptors of molecules to some type of activity (e.g., affinity to a given macromolecule); modeling this as a function of the descriptors:

$$a = f(\mathbf{d}) \tag{1}$$

where a is the activity of the compound, $\mathbf{d}$ is the descriptor vector of a compound, and f is a function.

By contrast, proteochemometrics takes all interacting entities in the system into account when modeling the interaction activity and can hence be viewed as a generalization of QSAR, as shown in the figure:

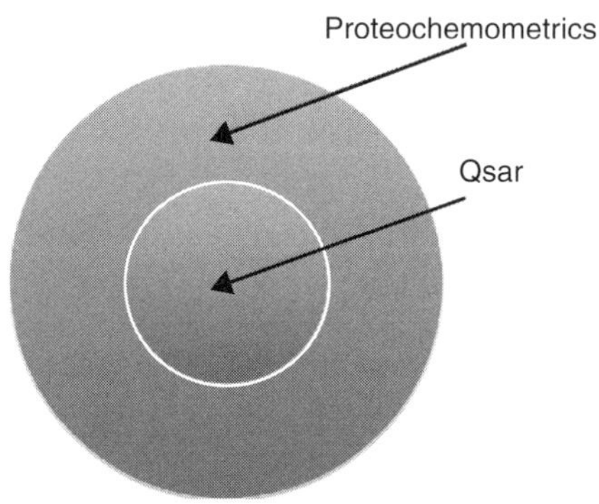

In the general case of proteochemometrics modeling, the interaction activity as a function of the descriptors of all interacting entities can be mathematically described as:

$$a = f(\mathbf{d}_1, \ldots, \mathbf{d}_n) \tag{2}$$

where $\mathbf{d}_1, \ldots, \mathbf{d}_n$ are the descriptor vectors related to interacting entity $1, \ldots, n$, respectively.

Proteochemometrics has generally been used to model the activity between a macromolecular unit (target) and a small organic molecule or peptide (ligand), in which case the proteochemometric equation (2) reduces to:

$$a = f(\mathbf{d}_T, \mathbf{d}_L) \tag{3}$$

where $\mathbf{d}_T$ and $\mathbf{d}_L$ are the target and the ligand descriptor vectors.

Both the QSAR and proteochemometric equations (1) and (2) can be extended to include descriptors of environmental conditions (e.g., pH, temperature, buffer composition) of a given interaction.

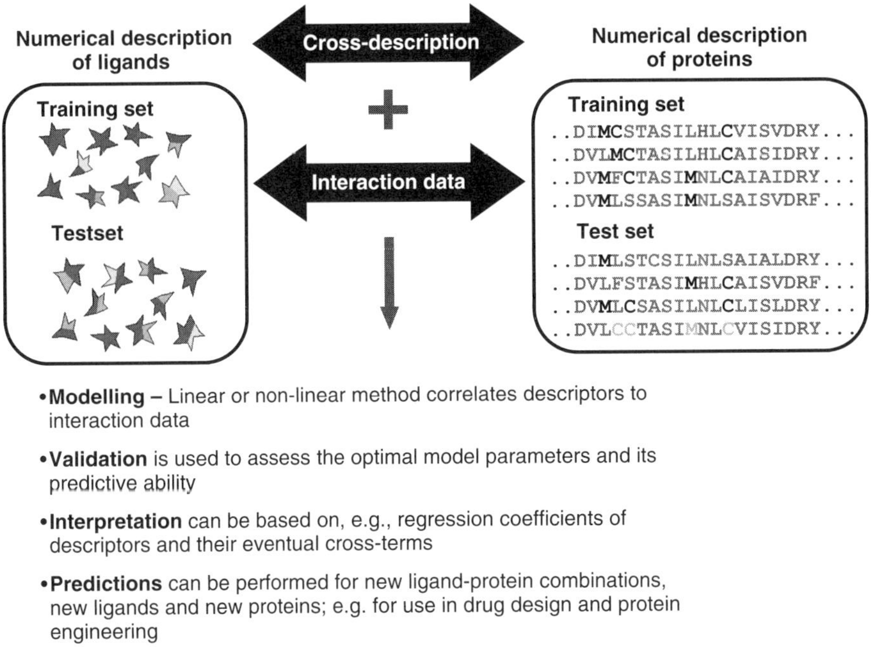

FIGURE 1 Overview of the principles of proteochemometric modeling. The numerical descriptions of ligands and proteins are related to the structural and chemical properties of the ligands and proteins studied. Cross-terms can be included to model the nonlinearities when a linear correlation method is used.

A typical task in proteochemometrics, the analysis of measured interaction data for series of proteins with series of small organic molecules, is outlined schematically in Figure 1. The interaction data for the training set proteins and ligands are correlated with descriptions of chemical properties of the protein–ligand pairs by using a suitable statistical method. Data for test set proteins and ligands are then used to validate the resulting model (i.e., to assess the model's predictive ability). Once the validity has been affirmed, the model can be used for interpretation of factors that determine protein–ligand interaction and for prediction of the activity of novel proteins and novel ligands.[5]

Correlation of the data [i.e., estimation of f in equation (2) of Box 1] can be performed using virtually any linear or nonlinear statistical learning technique. To this end, partial least-squares projection to latent structures (PLS) has been used extensively, due to its robustness and the interpretability of its resulting regression models. Other methods, such as ridge regression, rough set rules, support vector machines, and neural networks have also been used with success. The chemical description usually includes three blocks: ligand descriptors, protein descriptors, and ligand–protein cross-description (described in detail below). Interaction activity can be measured by any standard interaction or activity assay, such as radioligand binding or enzyme kinetic assay. Interaction data can be real-valued numbers, such as the logarithmically transformed ligand–protein dissociation or inhibition constant (pK_i), or on an ordinal scale (e.g., "no interaction," "weak interaction," "strong interaction," etc). (In fact, several proteochemometric studies applied successfully to the simple dichotomous case where the interaction data comprised a binary number presenting "interaction" or "no interaction.")

Data sets are often compiled from several sources where the assay conditions are not identical for all ligand–protein pairs, which can influence activity measurements. For example, the buffer compositions of assay systems are known to affect the interaction strength. In such cases, the differences between assays can be accounted for by complementing the ligand–protein description with assay environment descriptors, which are often shown to improve the model and contribute to the understanding of molecular interaction processes. In principle, any type of data of relevance to explaining the interaction of the entities studied can be added to a proteochemometric model, as it pertains to multivariate modeling in the general case.

Description of Interacting Molecules and Their Interaction

Organic Compounds Description of organic compounds is a well-developed field in QSAR. Compound properties can be represented at various levels of complexity. The simplest are binary representations, where binary numbers are assigned to the molecule depending on the presence or absence of particular features. The scope of any model based on binary descriptors becomes very limited, however as it does not allow one to interpolate or extrapolate between the binaries, and hence predictions for new molecules having features beyond those already existing in the dataset can't be done. A useful approach that remedies this deficiency is to use physicochemical properties, which include such descriptors as molecular weight, van der Waals volume, log P, and many others. When very many such descriptors are collected, they will relate to each other and be correlated. Calculating their principal properties can then be done, which results in sets of orthogonal descriptors. An advantageous approach is to describe the building blocks of organic molecules with such physicochemical property descriptors, as this leads to very fast computation of descriptors for combinatorial libraries.[14]

Structural descriptors based on the structure of the molecule are also in use and include one-dimensional (constitutional) descriptors such as count of atoms, bonds, rings or functional groups, two-dimensional (topological) descriptors, which include various connectivity indices calculated from a molecular graph, and three-dimensional descriptors, which are based on the 3D structure of the organic compound. In the latter case, the compound's 3D structure is generated using molecular or quantum mechanical approaches, for which 3D descriptors can be computed. These include, for example, molecular interaction field–based descriptors such as CoMFA,[15] CoMSIA,[16,17] and GRIND[18] descriptors. GRIND descriptors have been found particularly valuable in proteochemometrics, as they are independent of the rotation of the organic molecule.[4] This is because in many settings of proteochemometrics, the orientation of the molecule to its target is not known. In proteochemometrics modeling we are also dealing simultaneously with different targets where the orientation of the molecules binding may differ. As GRIND descriptors are based on distances between molecular interaction fields within the organic molecule itself, the orientation of the molecule becomes irrelevant to the descriptor representation. The GRIND descriptor concept was also recently augmented with a rotation-independent method to describe molecular shape, called the TIP field,[19] which has also been found useful in the description of organic molecules in proteochemometric modeling.[20]

In proteochemometrics, organic molecules may also be described by taking into account the presence of their multiple conformations and orientations, protonation states, and more. These types of representations of organic molecules are well known in the field of SAR, and the interested reader is referred to ref. 21 and 22 among other accounts, for more in-depth coverage of the subject.

Peptides and Proteins Proteins and other biomacromolecules can in principle be described in the same way as can any other chemical structures (e.g., as described above for organic molecules). However, because of their size, this leads to extreme complexity of the descriptions. In proteochemometrics a simple approach to remedy this is to focus on the polymeric nature of the macromolcules and describe them based on their building blocks: in the case of proteins, the amino acids. In fact, this should be sufficient, as the diversity of protein structures and functions is all a result of variations of chemical properties in the primary amino acid sequences, stated such that one also takes into account their eventual posttranslational modifications.

Accordingly, describing the properties of the amino acids becomes an important issue in proteochemometrics. Traditionally in bioinformatics, the letter codes of the amino acids have been used. In proteochemometrics this would correspond to using a vector of 20 binary variables taking into account natural amino acids only. Obviously, a more effective approach is to take into account the amino acids' chemical properties, as this would make generalizations into chemical effects caused by the amino acids possible. Using extensive chemical descriptions of the amino acids, as in QSAR (e.g., CoMFA, CoMSIA, GRIND), leads to very large arrays for describing even moderate-sized proteins, which would cause a problem in the following modeling steps. An approach that has been very successful in proteochemometrics is to use descriptors of principal physicochemical properties of the amino acids. One early study applied principal components analysis (PCA) to 29 measured and calculated physicochemical properties, transforming them into three uncorrelated (orthogonal) descriptors called *z-scales*.[23] These scales essentially summarize the hydrophobic, steric, and electronic properties of the amino acids. In a more recent study, *z-scales* were extended to characterize properties of both natural and synthetic amino acids and included five principal property vectors.[24] The *z-scales* have proven very useful in proteochemometrics and have now been applied in quite a number of studies. Other descriptors representing individual properties of amino acids have also been proposed. These include the amino acid flexibility and rigidity scales of Gottfries.[25] Another set of scales are the *tciz-scales* for peptide description, introduced by Muthas and co-workers in 2007.[26] These include scales for both α- and β-amino acids and are derived by computation only, the idea being that they can easily be extended to amino acids not covered by the present tciz scales, as well as being easy to interpret in terms of chemical modifications. This has a bearing in drug design when moving from a peptide to a nonpeptide.

With descriptors of amino acids in hand, it is straightforward to encode any protein by numbers representing the descriptors for the amino acid of each of its sequence positions. However, as most statistical modeling requires uniform matrices, it is necessary to align the proteins in a series prior to forming the descriptor matrix. This is straightforward when the proteins are related structurally. It is then reasonable to assume that comparisons of the numerical descriptions of the aligned sequences reflect the chemical variation between the proteins. Even with the presence of alignment gaps, this approach can be used, as the gaps can be treated as missing data or be represented by an indicator variable showing the presence or absence of amino acid in a particular position.

A large array of proteochemometric studies covering groups of alignable proteins is now available. However, generalization of proteochemometrics to allow modeling of broad groups of structurally nonrelated and nonalignable proteins is an interesting issue. From a theoretical point of view the ability of a molecule to interact with a protein binding site is solely dependent on the chemical properties of that binding site. Hence, obtaining a method for a generalized description of "all" binding sites may provide an opportunity for a proteochemometric model to cover the interaction of ligands with all types of binding sites. In fact, successful attempts in this direction has also been made in which the proteins are described by alignment-independent protein substructures[8,12]; very promising achievements along these lines are discussed further below. Studies of protein classification also suggest that sequence monomer-based description can allow broad comparisons of physicochemical, structural, and functional properties of proteins by applying auto-cross-covariance transformations,[4,27] Fourier transforms,[28] and other approaches[29,30] that normalize sequences of different length. However, thus far, none of the latter approaches have been used in proteochemometric modeling.

Description of Interaction Relating a combination of ligand descriptors to target descriptors using a linear multivariate approach would merely comprise a linear interpolation of the activities of the ligands for the targets and hence not reveal anything new compared to studying each target separately. It is well known that ligands may show high activity for some targets while being excluded from other targets. In pharmacology this is termed *selectivity* and is, of course, due to particularly favorable interactions of some targets with some ligands, something that is generally regarded to be a desirable feature for a drug. In proteochemometrics the "selectivity portion" can be viewed as the nonlinear part of the interaction, which is governed by selectivity-delineating properties of the targets and the ligands. In other words, the nonlinear portion reveals the differences in the chemical effects of the ligands and targets in combination that create the selectivity. In fact, this information can be used to localize where the interactions occur in chemical space. That this is so will become evident from the following discussion and from studying the example in Box 2.

Although a nonlinear regression approach could be used to reveal the nonlinear effect, linear approaches have been found very useful in proteochemometrics. In the latter case the nonlinear part can be described as a polynomial expansion of the linear terms by forming cross-terms of descriptors. Hence, using a linear method, the correlation of interaction activity y to L ligand descriptors, P protein descriptors, and their cross-terms is expressed by the regression equation

BOX 2: MODELING THE INTERACTION SPACE IN PROTEOCHEMOMETRICS

The interaction space is defined as the noncovalent interactions spanning all interacting entities in the sets of proteins or other target biomolecules and their ligands. The interaction space can be described from the chemical space of the interacting entities. Modeling the interaction space reveals the points of interaction of the entities studied. However, it can be revealed only when the interacting entities show selectivity within the sets of molecules being studied. In proteochemometrics a linear or nonlinear regression approach can be used to correlate molecular properties with interaction activities. When a linear approach is used, the nonlinear relation in the data can be revealed by extending the description of the chemical space of the interacting entities by forming cross-terms between descriptors. The regression coefficients of the molecular descriptors and the cross-terms formed are highly useful in an analysis of the interaction space, as is evident from the simple example in the figure.

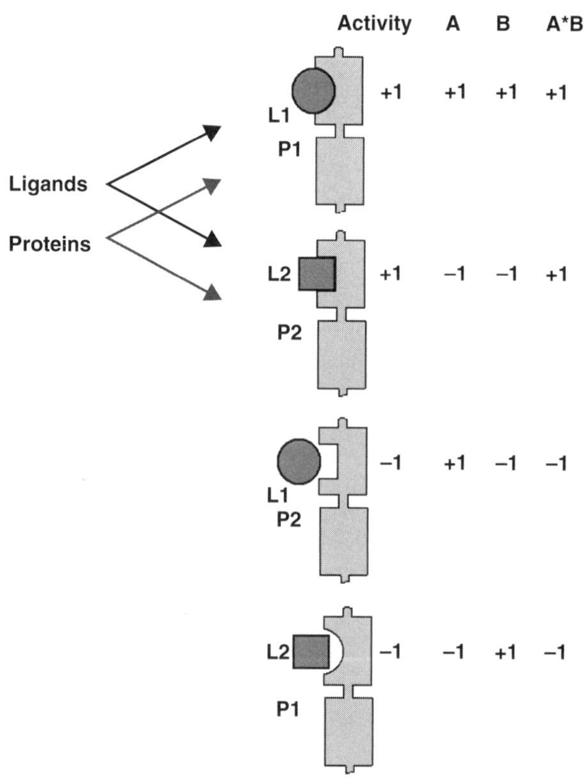

In the example shown, two ligands, L1 and L2, interacting with two proteins, P1 and P2, differ in some structural property, represented by the descriptor A. This descriptor is assigned the value 1 for L1 and −1 for L2. In a similar way, imagine that the proteins differ by some property that is represented by descriptor B, attaining values 1 for P1 and −1 for P2. Let us assume that the ligands have opposite selectivity. Thus, L1 binds with high activity to P1 and with low activity to P2, whereas L2 binds with high activity to P2 and with low activity to P1. Accordingly, the activity can be represented by the numbers 1 and −1. Inspection of the figure shows that there is no correlation between A and the activity (since both ligands have the same average activity for the two proteins). Nor is there a correlation between B and the activity, since the proteins do not differ in the ligands' average activity. By contrast, a cross-descriptor between A and B shows a one-to-one correlation with the interaction activity. Hence, one may conclude that the ligand property A and protein property B cooperate in creating selective interactions (i.e., we have localized their interaction in chemical space).

If descriptors A and B and their cross-term A × B would be correlated with the interaction activity by some linear regression approach, A and B (reflecting differences in the average affinity of the ligands and proteins, respectively) would obtain regression coefficients 0 and their cross-term (reflecting differences in selectivity) would obtain coefficient 1. Hence the coefficients are useful to interpret where interactions between the ligands and targets occur in chemical space.

$$y = \overline{y} + \sum_{l=1}^{L} (\text{Coeff}_l \times x_l) + \sum_{p=1}^{P} (\text{Coeff}_p \times x_p) + \sum_{l=1,p=1}^{L \times P} (\text{Coeff}_{l,r} \times x_l \times x_p) \tag{4}$$

where the Coeff's are the regression coefficients, which by their sign and magnitude represent how the descriptors are correlated to interaction activity y.

Analysis of equation (4) reveals the role of each block of descriptors in explaining differences in protein–ligand activity. If we imagine a hypothetical case where the activity can be explained solely by ligand descriptors (i.e., ligand descriptors were the only ones obtaining nonzero coefficient values), this is possible only if any one ligand has identical activity for all proteins. If the activity were shown not to be affected by properties of the target proteins, a reasonable interpretation would then be that the ligands interacted only with the invariant parts of the targets in the studies series. In reality, however, the activity of any ligand for different protein targets varies, which in the regression model can then be explained partially by the protein descriptor block. However, a model including only ligand and protein descriptor blocks is not able to explain differences in the ligands' selectivity profiles nor in the proteins' selectivity profiles. Moreover, such a model would predict that any change in a ligand would result in an equal increase or decrease in the activity for all proteins, placing us in the same unlikely situation that ligands interact only with the invariant parts of proteins. (A similar interpretation from the protein point of view would suggest that the ligands also interacted with proteins by virtue of some invariant ligand feature). Thus, even if the model could partially explain activity differences, any interpretations would lead to logical flaws, and no predictions as to how to alter the selectivity would be possible.

The situation changes by introducing cross-terms, which represent the complementarity of ligand and protein properties. This is illustrated in a simplified form in Box 2, which includes a model for the interactions of two proteins with two ligands showing opposite selectivity. Using real numbers rather than binary descriptors allows one to describe the studied effects on a continuous scale rather than in all-or-nothing fashion. Using richer descriptions will thus make possible a more detailed interpretation of the nature of the ligand–protein interactions. Thus, for a real data set with numerous structural or physicochemical descriptors, analysis of regression coefficients for ligand–protein cross-terms allows us to localize the place and type of molecular interactions.

Interaction effects can be studied inside the ligand or protein target as well. In an fashion analogous to forming ligand–target cross-terms, this can be done by forming intraligand or intratarget cross-term blocks. These additional cross-terms are useful in situations when ligand and protein properties cooperate within the respective entity, causing intramolecular effects (e.g., these can correspond to intramolecular interactions and conformational changes within the series of molecular entities being studied).

3 EXAMPLES OF THE APPLICATION OF PROTEOCHEMOMETRICS

Next we provide examples of studies applying proteochemometrics and indicate pertinent results obtained by use of the technology.

Modeling the Interaction of Oxazoles with α_1-Adrenergic Receptors The first study introducing proteochemometrics was conducted using three native and 15 chimeric and point-mutated α_1 adrenergic receptors, interacting with a series of 12 organic compounds with the same generic structure (i.e., comprising 4-piperidyl oxazoles.[31]) The varying sequence residues of the receptors were described by *z-scales* and the substituents in three positions of 4-piperidyl oxazoles were described by binary descriptors. The descriptors and their cross-terms were correlated with the binding affinity using PLS. The model obtained showed excellent statistical performance and revealed that two particular amino acids, Ile85 and Val185, determined the overall high affinity of compounds for the ADA1A receptor. It also showed that interactions with Phe86 discriminated between ADA1A selective and nonselective compounds. The latter conclusion could be drawn from the high level of importance of cross-terms formed by descriptors of Phe86 with compound descriptors. Equal importance was given to several intraprotein cross-terms, indicating the presence of distant interactions in the receptor, which, for example, could affect packing or tilting of the transmembrane regions. This study was important, as it showed for the first time the feasibility of using proteochemometrics and that it could be used to localize the chemical effects responsible for drug targets' interactions with organic compounds.

Modeling Ligand Interactions with Melanocortin Receptors A series of subsequent studies was devoted to melanocortin (MC) receptors. These are G-protein-coupled receptors for which natural ligands include melanocyte-stimulating hormone (MSH) peptides. There exist four subtypes of MSH-binding MC receptors: MC_1, MC_3, MC_4, and MC_5R. In order to create chemical variation among the receptors, a receptor library was genetically engineered by dividing the receptors into three sequence stretches and then shuffling them so as to create chimeric receptors with sequences from three of the four receptor subtypes. Using this approach in conjunction with D-optimal experimental design, to assure selection of as a diverse library as possible while having to manufacture as few entities as possible, led to a library of 14 multiple chimeric and 4 wild-type MCRs, covering well the chemical space of the MCRs.[7]

One study using this library focused on validating the ability of proteochemometrics to localize interaction of the receptors' amino acids with MSH peptides. To this end, interaction with the MCR library of the natural α-MSH hormone and two MSH peptide analog with differing receptor selectivity was assessed and the data were subjected to proteochemometric modeling. The model gained showed excellent performance characteristics and allowed us to localize amino acids involved in creating MC_1R selectivity for the α-MSH peptide. The model thus identified seven amino acids located in or around the vicinity of the MSH peptide-binding pocket responsible for this. Interestingly, sequence stretches in the third intracellular loop at a distance of more than 20 Å from the binding site of α-MSH also proved to be important. All the predictions obtained using the proteochemometric model were verified posteriori by site-directed mutagenesis. This study is of very great importance, as it validated for the first time the ability of proteochemometrics to derive correct predictions of both short- and long-range effects in proteins related to the binding of a ligand to its binding site. One can envisage that these effects arise due to both direct interactions of receptor amino acids with the MSH peptide as well as to effects arising at a distant site in the protein that affects the mobility of the receptor or the conformation of the receptor's ligand-binding pocket.[7]

Another study using the MCR library was directed at modeling the interaction of a diverse set of organic compounds with MCRs.[32] Also in this case, an excellent model was created that could pinpoint amino acids in the receptors involved in interaction with the organic compounds. Amino acids that presumably afforded direct points of interaction at the binding site with the compound, as well as selectivity forming amino acids along the path for entrance of the compound into its binding pocket, were delineated from the model. Moreover, the model was verified to provide a priori correct predictions of the binding strength of novel organic compounds for the MCRs.[32] Several similar studies successfully modeled the interaction of linear Trp9 substituted[33] or cyclic MC_4R selective peptides[34] with the MCRs.

On top of that, one study compared proteochemometrics with QSAR, applying the computations onto only the four MSH-binding MCRs interacting with 54 structurally diverse organic compounds.[13] Despite the small number of targets, the study showed that proteochemometrics resulted in more stable and interpretable models than separate QSAR models made for each MCR subtype.

Modeling of the Interaction of Organic Amines with Amine GPCRs A disadvantage of proteochemometrics seemed initially to be that it required mutated targets or chimeric constructs. However, many target proteins exist in large structurally related families such as for the amine G-protein-coupled receptors (GPCRs). Two studies were devoted to exploring the use of wild-type amine GPCRs only, and in both cases good interpretable proteochemometric models were obtained.[20,35] The more recent of the two studies modeled the interactions of a diverse set of tricyclic and piperidine/piperazine ring-containing amines with representatives of all amine GPCR families: namely, the adrenergic, serotonin, dopamine, histamine, and muscarin receptors; in a total of 31 amine GPCRs.[20] In this study the organic compounds were characterized using rotation- and superimposition-independent GRIND descriptors, and the alignable transmembrane segments of the receptors were represented by *z-scales*. The large size of the descriptor matrices required preprocessing with principal components analysis prior to forming ligand–target cross-terms to obtain a data matrix of surmountable size; a good proteochemometric model was then obtained which was predictive of the interaction strength of an external set of organic amines for amine GPCRs. The model was even predictive for new amine GPCRs, even for the case when both the GPCR and the organic compound were a priori unseen by the model. Moreover, using GRIND descriptions of the organic molecules for the model allowed detailed interpretations of the structural and chemical features, making them selective for one or the other type of amine GPCR.

Accordingly, the proteochemometrics approach in drug design holds great promise for amine GPCRs; the models formed by the approach can be interpreted chemically so as to rationally form improved compounds, or they could be used for rapid screening of virtual chemical libraries to sort out useful novel compounds.

Models Exploiting 3D Structural Data of Proteins All of the studies reviewed above obtained their target protein descriptions from the targets' primary amino acid sequences. Another interesting trend is to derive these descriptors from the protein's 3D structures. Two such studies used a novel approach in which the target proteins are represented by descriptors of commonly reccurring local substructures of their 3D structure (see ref. 36), which leads to general descriptions of the proteins without the need to align them structurally.[8,12] While the first of these studies was a pilot study covering only hydrolases and lyases to show the feasibility of the approach, the latter was directed to providing a proteochemometric model covering all six major enzyme EC classes. The training set in the latter study consisted of 826 co-crystallized enzyme–ligand complexes with examples covering all EC1 to EC6 classes, while the ligands, which represented a diverse set of compounds ranging from natural substrates and products, cofactors, and activating compounds to synthetic inhibitors, were characterized by 27 structural and physicochemical property descriptors. A nonlinear model was fit by applying a support vector machine with a radial basis kernel. The resulting model was validated on an independent testset comprising 542 enzyme–ligand pairs that were not co-crystallized and hence represented cases where the binding mode was unknown. The model obtained showed good predictive ability, as indicated by results of cross-validation as well as by predictions for the external test set.[12] Thus, although the locations of the binding pockets in the test-set enzymes were not specified and the exact geometries of bound ligands were unknown, the model could provide equally good predictions for noncrystallized as for the crystallized entities. These results are very encouraging in view of the ability of the model to give accurate predictions over a range of binding affinity (pK_i) values spanning as much as 0.7 to 11. It is noteworthy that predictions could be made covering structurally very different and essentially unalignable proteins where more than half of test-set enzymes shared less than 30% sequence identity with the nearest training-set enzyme. Thus, the approach seems to hold great promise for obtaining large-scale proteochemometric models covering a large portion of the proteomes.

Another attempt using descriptors derived from 3D structures of protein–ligand complexes, which corresponds to a proteochemometrics approach was reported by Lindström et al.[37] In this study, proteins were (among other types of descriptors) represented by auto-cross-covariances of coordinates of α-carbon atoms and auto-cross-covariances of elements of secondary structures, but the models had a more limited scope and were less predictive than those presented by Strömbergsson et al.[12] Yet another 3D approach, also utilizing proteochemometrics, represented ligand–protein interfaces by quadruplets of nearest atoms.[38] The proteins were described in a very simple manner by the presence or absence of quadruplets with distinct combinations of atom types (C, N, O, S, halogens and P, and others) (i.e., binary descriptions were used), which, interestingly, proved to be sufficient for obtaining predictive models.

Although the present requirement for accurate x-ray structures of ligand–protein complexes places some limitations on all these approaches, results are very encouraging; in particular, the approach of Strömbergsson et al.[12] is promising, as it is the most general and provides very wide models over the proteome. Further studies along these lines are thus warranted.

Exploring the Mutational Space of Retroviral Proteases Another line of research in proteochemometrics was to take the advantage of the high mutational rate of retroviral proteases. Only for the HIV protease, which comprises a dimer, with each monomer holding only 99 amino acids, 24,000 mutant variants have been described comprising single-point mutations or combinations of mutations in more than 80 positions. The appearance of mutations is a clinical problem in HIV, as it leads to the development of drug resistance in treatment with antivirals.

In a first attempt we modeled the cleavage rate of natural and synthetic cleavage sites for the mutant variants of the HIV protease, including drug-resistant variants. We soon found out that including more information by incorporating data for other retroviral proteases as well, improved the models markedly. Accordingly, we set out to collect essentially all publicly available data for retroviral proteases; the final data set, representing essentially all data from 16 years of retroviral research, included 61 retroviral proteases represented among nine different retroviral species, with a total of over 2000 protease–cleavage site combinations.[10] Two proteochemometric models were made; one for cleavage/no cleavage (i.e., a binary representation of interaction activity) and one for the peptide cleavage rate, k_{cat}/K_m. In both models, amino acids in the proteases and the peptides were represented by *z-scales*. Both models showed excellent performance and were validated experimentally by first screening a large peptide library virtually, and, based on the results, randomly selecting 30 peptides which were subsequently synthesized and assayed on drug-sensitive and drug-resistant mutated HIV-1 proteases. Again these models proved to perform excellently. For the cleavage model, the a priori prediction accuracy proved to be 100%, and for the cleavage rate model, predictions showed 68% accuracy within a k_{cat}/K_m value of 1 log unit. Using these models, detailed interaction maps for the roles of amino acids as well as for the chemical properties and cooperative effects in the substrates of importance for substrate cleavage could be made[10] (Figure 2).

Yet another study used proteochemometrics to model the susceptibility of HIV protease inhibitors used clinically.[39] Also in this case, a highly predictive and interpretable model was obtained. A server based on the model that can predict the susceptibility for protease inhibitors from the HIV protease sequence is available at http://hivdrc.org.[40]

The success of proteochemometrics to model resistance in HIV indicates that the approach will find important uses in designing novel improved retardants in HIV that can combat drug-resistant HIV.

Other Proteochemometric Studies Several other studies that used proteochemometrics successfully include the modeling of the inhibition of cytochrome P450 enzymes by organic compounds[41] and dengue proteases,[42] and of antigen–antibody interactions.[9] Other studies using the proteochemometrics approach modeled the interaction of PDZ domains with peptides[43] and ligand interactions with the entire superfamily of GPCRs.[44]

Moreover, qualitative protein–protein interaction models (i.e., interacting/not interacting) have been created intended for whole proteome

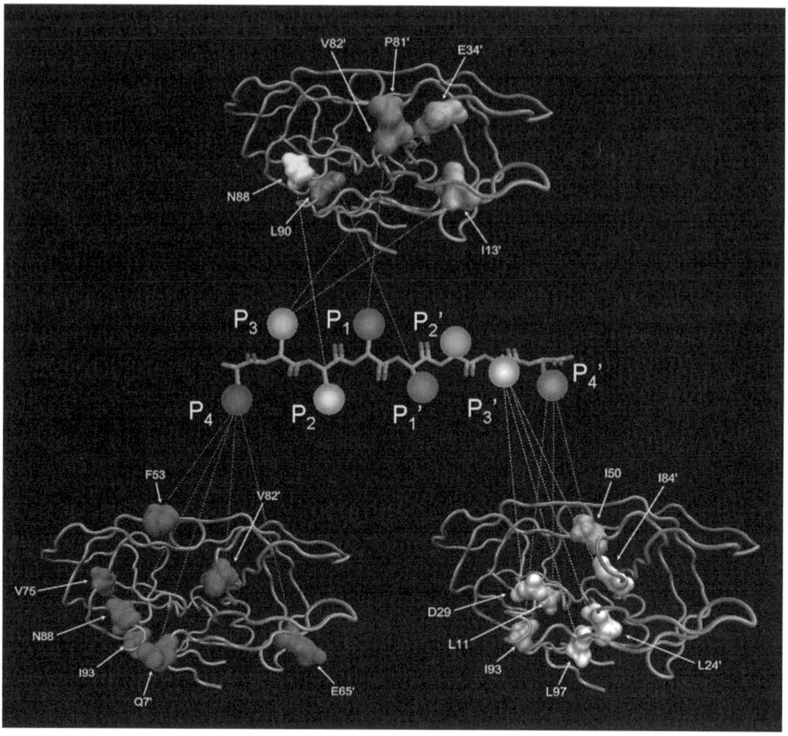

FIGURE 2 Cross-dependencies of amino acids in retroviral proteases and their substrates identified from a proteochemometric model for substrate cleavage, made on 61 wild-type and drug-resistant mutant forms of proteases from nine retroviral species. Shown are the positions for amino acid pairs in the P4–P4′ positions in the substrates whose regression coefficient sizes were of significant magnitude in the proteochemometric regression model. Note that the protease is represented as a "cartoon" on the HBX2 HIV-1 protease 3D structure; the underlying proteochemometric model did not use any 3D information. In essence the cross-dependencies give information on direct and indirect interactions between substrates and protease amino acids that determine the rate of substrate cleavage. (Modified from ref. 10, courtesy Aleksejs Kontijevskis.)

interaction mining.[45,46] These are interesting studies in which the models show very high precision and sensitivity when validated to data for related organisms.

4 SOFTWARE FOR PROTEOCHEMOMETRICS

Performing proteochemometrics modeling requires handling of heterogeneous chemical and biological data using a plethora of tools for information management, processing, and data analysis. Each step in the proteochemometrics workflow has traditionally been performed using separate software packages joined in an ad hoc fashion. This requires not only domain knowledge but also expertise in computer science. Figure 3 shows an example for a typical proteochemometrics analysis: (1) literature or database searches to obtain or verify interaction data (often available only in text or tables in scientific publications); (2) search and retrieval of sequences from public repositories; (3) identification of chemical structures from databases or from images in scientific articles with manual redrawing; (4) sequence alignment using suitable software or services; (5) alignment editing; (6) sequence descriptor calculations (which are usually done with ad hoc solutions such as using a Perl script or an Excel macro); (7) processing of chemical structures (e.g., converting 2D structures to 3D structures); (8) calculation of chemical descriptors with commercial software; (9) manual setup of data sets in a spreadsheet; (10) statistical analysis in appropriate software; and (11) interpretation of the results from the analysis and the tracing back from analysis results to biological and chemical insights.

Examples of packages that can be used for some of the steps in a proteochemometrics analysis include software for chemical structure management and processing, such as MOE,[47] Sybyl,[48] Sarchitect,[49] and Discovery Studio.[50] Dedicated packages, such as Dragon[51] and Almond[52] for descriptor calculations and SIMCA[53] and Unscrambler[54] for statistical analysis, have been used extensively in proteochemometrics. All these programs are available only in closed-source commercial software and in many cases are quite expensive, which makes them not very cost-effective solutions in proteochemometric modeling. Further, their focus is on chemoinformatics, not really on statistical analysis, which makes them quite limited in their statistical offerings. It could thus make sense to use software developed for statistics, such as SPSS[55] or SAS,[56] to perform the data analysis. However, these packages focus on statistics, not on chemo-bioinformatics, which makes it tedious to use them to manage chemical and biological information; the bottom line is that they are expensive and not well tailored for the tasks of proteochemometrics. Open-source solutions and services can be used for some tasks in proteochemometrics, but the software available is, as a rule, developed to perform one single task, and most is equipped with very limited user interfaces; an example is JalView[57,58] for sequence alignments. R[59] is an excellent free and open-source language and environment for statistical computing. However, R does not handle chemo-bioinformatics tasks, and it lacks a well-developed graphical user interface, which makes it difficult to use for the inexperienced user.

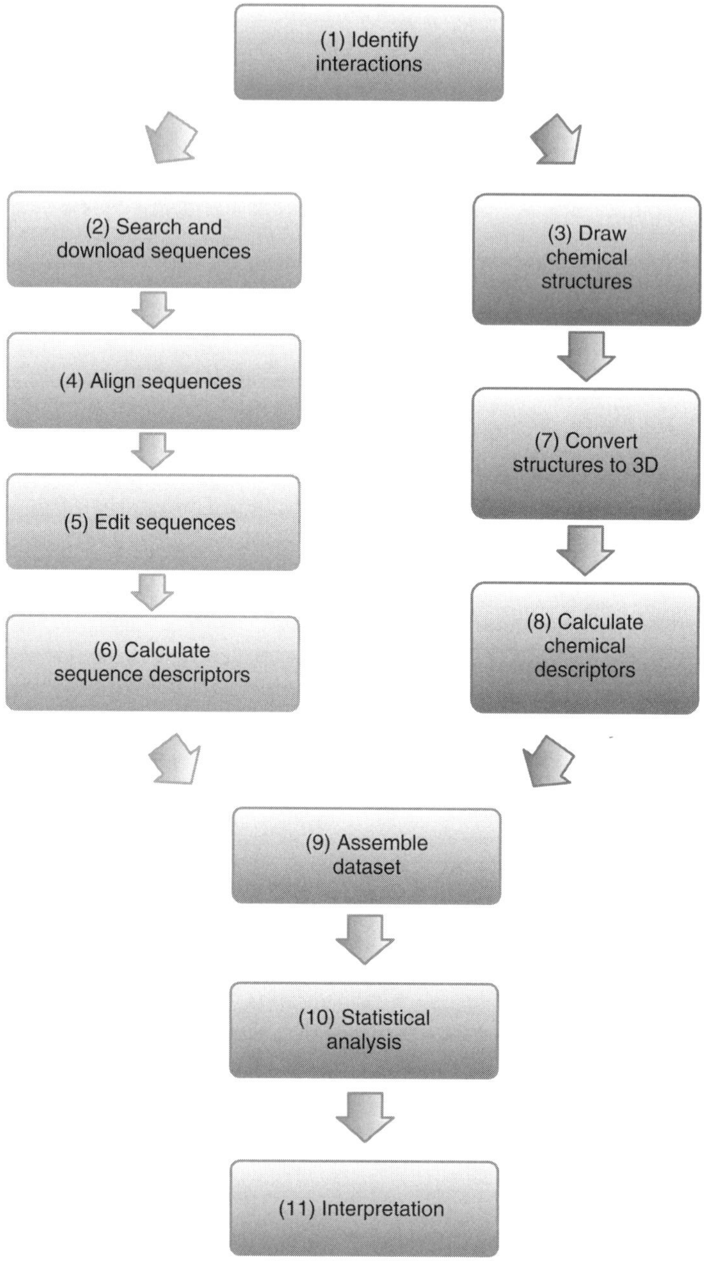
(1) Identify interactions
(2) Search and download sequences
(3) Draw chemical structures
(4) Align sequences
(5) Edit sequences
(6) Calculate sequence descriptors
(7) Convert structures to 3D
(8) Calculate chemical descriptors
(9) Assemble dataset
(10) Statistical analysis
(11) Interpretation

The drawbacks to the overall procedure for proteochemometrics modeling using the solutions noted above are the low throughput and the difficulty, or even lack of possibility, of validating (since the steps are performed independent of each other and hence cannot be validated jointly). Further, the many manual steps involved, such as data formatting and conversion, make the workflow error-prone and the final results complicated to interpret because of the difficulty in relating statistical analysis output to the original chemically derived input.

Bioclipse The requirement for many different incompatible software tools and the manual conversion steps made it in practical to carry out proteochemometrics analyses in a high-throughput fashion. A new platform was needed. This prompted the start of an open-source project in 2005 which has led to a feature-rich workbench for life science called Bioclipse.[60,61] Bioclipse is a "Rich client," which means it is a downloadable software application with a rich graphical user interface that takes full advantage of the local desktop computer but also makes use of networked resources such as databases and Web services[62,63] and computational resources such as grids and clusters. This makes Bioclipse an ideal platform not only for proteochemometrics but also for any chemo-bioinformatics task that requires massive analysis of complex biological

←

FIGURE 3 Proteochemometrics workflow. Each step in the example was performed using separate software packages joined by ad hoc programmed scripts. Using a series of amine GCPRs with representatives from all major known functional classes, the proteochemometrics analysis comprised the following steps: (1) Interaction data were retrieved from one database [the Psychoactive Drug Screening Program (PDSP) database[71]] and (2) protein sequences from another (Swiss-Prot database[72]). (3) The protein sequences were subsequently aligned using a GPCR-specific alignment algorithm,[73] (5) further curated in the JalView alignment editor,[58] and (6) described with the z-scales of Sandberg et al.[24] using a custom Perl[74] script. (4) Structures of compounds were drawn using ISIS/Draw.[75] (7) The structures were converted to 3D by the Corina unit of the Tsar 3.3[50] software package, partial atomic charges were derived using the Charge2 unit of Tsar 3.3, and their geometry was optimized by performing energy minimization using the Cosmic utility of Tsar 3.3. The 3D structures thus obtained were (8) described by grid-independent descriptors (GRINDs)[18] calculated by Almond 3.1[52] software. (9) A data set was then compiled using a Perl script in which the protein and ligand descriptors were concatenated and matched against their activity value. (10) The descriptor matrices derived from the small compounds and the proteins, respectively, were dimension reduced using the PCA implementation in SIMCA.[53] The principal components of the compound and protein descriptor matrices were then correlated with their interaction activity by means of PLS, as implemented in SIMCA (11), and the interpretation of the model was conducted manually.

and chemical data. Bioclipse has an open-source license, which means that anyone can download the source code and make changes. This promotes global collaborative development efforts as well as providing quick and responsive bug fixing. Bioclipse incorporates many features in bioinformatics, chemoinformatics, spectrum analysis, QSAR analysis, statistical modeling, statistical learning, the BioMoby system for biological services, and others. The ability to integrate data using the latest standards and to use local applications and remote services for chemical information management, advanced data analysis, and visualizations in 2D and 3D[60,61] from within the same application has met with great approval among the scientific community (see Figure 4 for an example).

Bioclipse is built on Eclipse,[64] which is a universal tool for constructing software applications. Eclipse is a general framework with an advanced component-based architecture that makes it simple to extend. All code is split up into plug-ins—even the core modules. A *plug-in* is a collection of functionality (Java classes) that can be integrated seamlessly with other plug-ins, such as algorithms, visualizations, and menu options. This allows components to be used as building blocks. In Bioclipse it is straightforward to extend almost every part, such as views, editors, wizards, menus, algorithms, and visualizations. Bioclipse defines extension points for bio- and chemoinformatics: for example, for adding new biological and chemical entities (e.g., sequence, protein, structure, scripts), new file formats

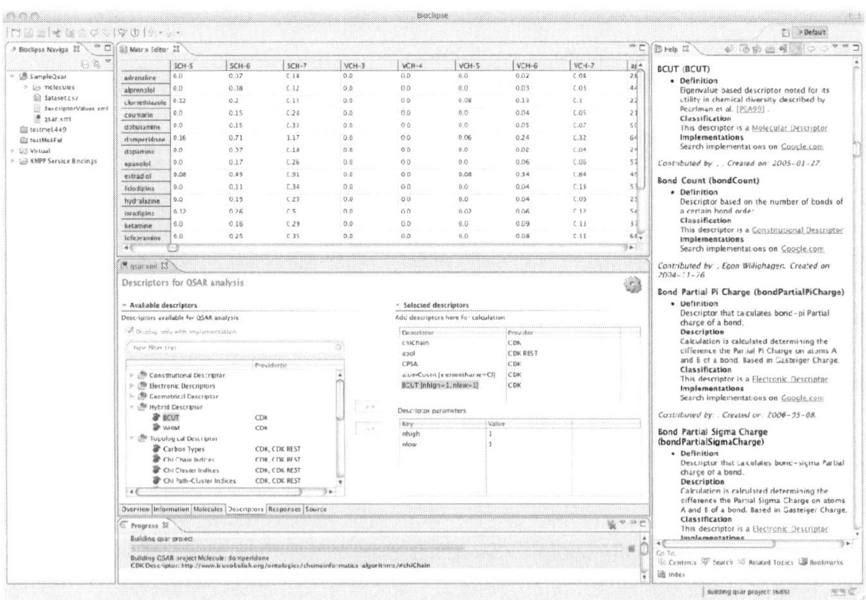

FIGURE 4 Screenshot from a QSAR experiment in Bioclipse.

and parsers, algorithm implementations, data analysis, and much more. End users can select desired features from all the Bioclipse plug-ins, and use them integrated from an intuitive, responsive graphical workbench. In this way it is easy to install only the features needed, and not clutter the workbench with undesired options.

For the developer, Bioclipse is the ideal framework for integrating new components and taking advantage of already existing building blocks. This permits the developers to focus on the actual scientific problem without worrying about the plumbing. Bioclipse is a *platform*, meaning that it is open-ended and sets no boundaries for extensions. Plug-ins from other Eclipse-based applications can be installed and run effectively within the Bioclipse workbench with few or no modifications, which gives Bioclipse such advanced features as version control, GRID access, online updates, internationalization, a searchable help system, and advanced XML processing.

Bioclipse-P Bioclipse-P is a Bioclipse plugin for experimental design and statistical learning tailored for proteochemometrics analyses and other demanding tasks in chemo-bioinformatics in development. Bioclipse-P takes its stance in a list of requirements compiled with the specific problems in mind that are encountered when designing experiments and analyzing data in proteochemometrics (and, in fact, within the biosciences in general). Some of these requirements are experimental designs that are applicable even when the design space is irregular, partitioning of data observations and variables into logical groups, assumption testing and outlier detection, handling of missing data and observation replicates, sound ways of dealing with ill-posed matrices and the problem of having many variables in comparison to the number of observations available, sturdy validation methods, and ease of model interpretation. Bioclipse-P is designed to be easily extensible and, accordingly, defines extension points that need to be implemented in order to add functionality. It is also bridged to the statistical programming language R; the power and immense functionality of R can thus be accessed from P.

Bioclipse-P and is presently available in a development version.[65] Linking Bioclipse and with Bioclipse-P will ultimately bridge chemoinformatics and extensive statistical functionality via an intuitive graphical interface and enable one to relate the output of statistical computations back to the biological and chemical input. A screenshot of the Bioclipse P prototype is shown in Figure 5.

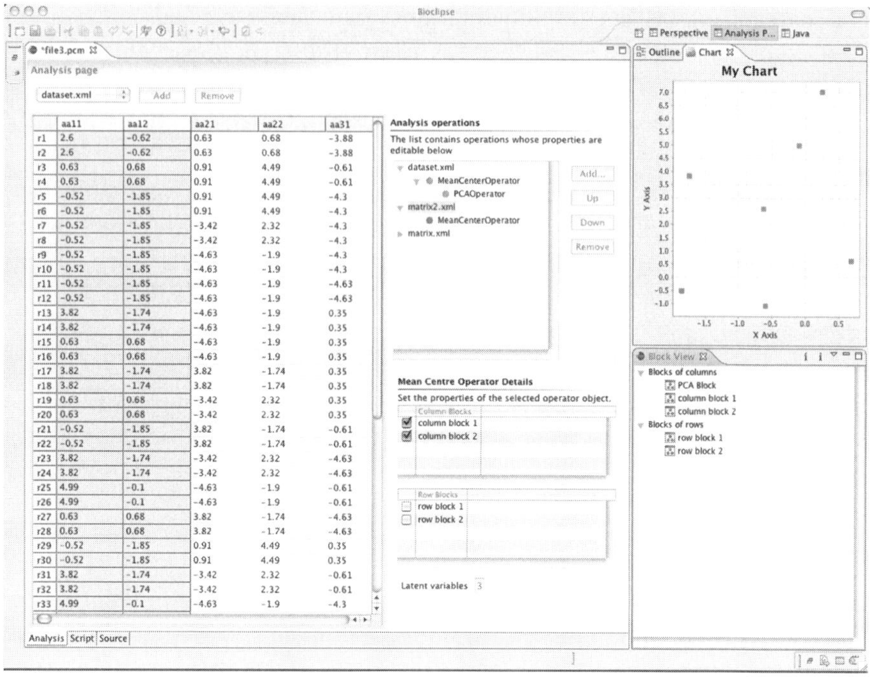

FIGURE 5 Screenshot of the Bioclipse-P prototype in development. The figure shows the matrix view (the spreadsheet functionality) to the left, the data analysis operations workflow in the middle, and the plotting view to the right.

Summarizing Software for Proteochemometrics The integrated environment of Bioclipse makes it an ideal platform for proteochemometric modeling. Bioclipse contrasts markedly to the Web-based initiatives developed within many other bioinformatics projects, which generally lack the flexibility and the intuitive type of user interface of Bioclipse. For the end user this means that in Bioclipse all functionality is available from a single desktop application that has the look and feel of the user's operating system (i.e., if run on Windows it looks like a Windows application). Bioclipse, being developed in the platform-independent programming language Java, runs on a virtual machine that is freely available for most operating systems, such as Windows, Linux, and Mac OS. The integrated approach, where all functionality available from within a single interface is possible with Bioclipse, contrasts fundamentally with the loosely coupled, sequential workflow that can be performed using commercially available software (cf. Figures 3 and 6). Bioclipse thus leads to increased productivity and fewer errors.

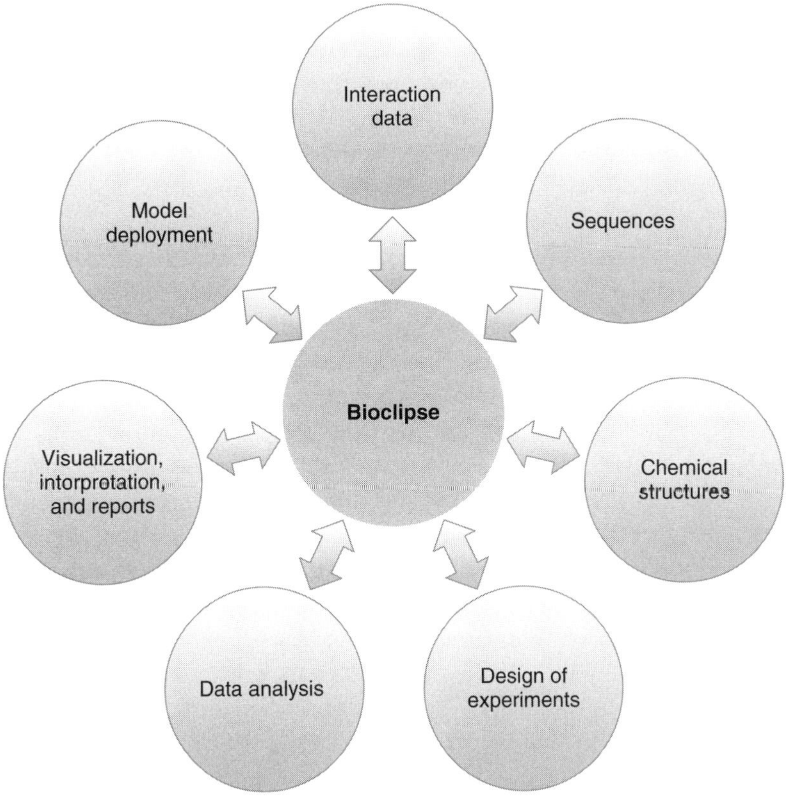

FIGURE 6 Use of the single application platform Bioclipse to integrate the components needed to perform a chemo-bioinformatics task, such as proteochemometrics. This contrasts markedly with the sequential and loosely coupled workflow shown in Figure 3 and enables proteochemometrics to be conducted on a larger scale and with a higher throughput. To exemplify the difference from proteochemometrics protocols used previously, we can compare the workflow depicted here with that in Figure 3. According to the new model, steps (1), (2), and (5) are performed with a single Web search from within the Bioclipse browser where all known interactions between organic amines and amine GCPRs are (due to Bioclipse's chemical awareness) retrieved in a compatible file format. Steps (3), (7), (8), and (9) may then be performed by accessing Web services or Bioclipse plug-ins, whereas steps (10) and (12) may be conducted using Bioclipse data analysis tools or, if necessary, by accessing high-performance computing facilities from Bioclipse.

5 THE BOTTOM LINE AND OUTLOOK FOR THE FUTURE

As we have discussed herein, incorporating information from many targets into a single joint model increases the accuracy of predictions for any single target being in scope for the model. While the first attempts to perform proteochemometric modeling were directed to closely similar targets interacting with a series of molecules having simple substitutions on a single scaffold, the developments have over time shown that predictive models can be obtained over target series where its members are lacking sequence and fold homology and are interacting with very diverse types of ligands, such as natural substrates and products, cofactors, and synthetic compounds.[8,12] Moreover, experimental validations have shown that proteochemometric models take into account not only chemical effects arising from close interactions of a ligand with amino acids at its binding site, but also distant effects causing indirect conformational and dynamic effects on the interacting entities.[7] In fact, such effects have been difficult to model with other technologies, including even with the use of extensive, costly, and resource-demanding molecular dynamics computations.

A natural question then is: How can it be possible to afford such broad and far-reaching predictable models based on comparatively simple chemical descriptions of the structural and chemical properties of the targets and their ligands? The answer is simple—it is because targets and ligands of such different types share enough chemical similarity to allow empirical modeling to generalize into predictive models over these broad domains. In fact, noncovalent interactions of molecules are governed by the same chemical principles: namely, ionic bonds, hydrogen bonds, hydrophobic interactions, and van der Waals forces. The ability to afford generalizations is then dependent only on how we describe the interacting entities in a proper chemical sense so that such generalizations are possible, even in between very different types of targets, irrespective of to which "biological class" or "structural class" they belong.

The results summarized in this chapter have immense impact in drug development and biological and chemical research. They mean that a systematic reuse of old information will help any development project aiming for something new (e.g., a drug for a new target). Thus, for any new development there would always exist prior information; the more effectively and systematically all relevant prior information is utilized, the less is the overall cost for the new development and the higher the probability of bringing it to final success. It also means that drug developments should be regarded as a parallel processes where the accumulating results from all different projects are systematically cross-utilized and fed iteratively into each other, rather than as each being conducted as a solitary process where all steps are conducted independently in a predetermined order along a single path.

It also means that the value of "old data" increases tremendously. Thus, even data of seemingly remote relatedness to a particular problem at hand can give some information of importance to it. However, unfortunately, today we mistreat data quite badly. Scientific journals do not allow raw data to be published, due to "space constraints," and scientific traditions force us to present data transformed into tables, graphs, or images, where much of the original information may have been removed or even become distorted. In proteochemometric modeling the original raw data are actually more useful, and negative data are of equal importance. In fact, an enormous amount of such data is confined in the private collections of researchers with little chance to reach the public domain. The information that two entities are *not* interacting has a lot of value, but such information is not easily published. Also, while genomics sequence data for proteins and structures for chemical compounds are fairly well organized in databases, their functional annotation is lagging behind. Interaction data are collected in a more or less random fashion and are made available in scientific publications, patents, proprietary collections among pharmaceutical companies, and to some extent in proprietary and public repositories, albeit sorted in nonuniform ways. Information concerning environmental factors of importance, such as incubation times, buffer compositions, and reactant concentrations, are often lacking. Accordingly, the systematic collection of data, the development of standards and software for retrieving the data, and the formation of the organizational structures to maintain the availability of such data is of immense importance.

Improvement of the technology as to how to make the descriptions of interactions in proteochemometrics would lead to increased predictive ability and reduce the requirement for examples in affording predictive proteochemometric models. One aspect here that should be incorporated thoroughly in future studies is the systematic application of experimental design technologies in the selection of targets and ligands in order to reduce the number of experimental tests required. It is well known that the use of statistical molecular design leads to a larger information gain at less cost (shorter time, fewer number of experiments, etc.) and to better predictive ability for the resulting models. It is important to point out that current "large-scale" proteochemometric models[8,12] may be useful for initial proteome-wide screenings, but they still lack the resolution required for detailed interpretation, design, and optimization of ligands for dedicated targets. Hence, continued research in the field of proteochemometrics seems of great importance. Development of visualization tools and other tools to aid drug design from high-resolution proteochemometric models is another line of research of great importance.

Proteochemometrics also makes it possible to afford a straightforward ligand design concomitantly over many targets. This is because a proteochemometric model encompasses all the various targets simultaneously and affording designs for one or several target amounts to the same.[11]

Although the mainstream in pharmaceutical development has for some time been to afford compounds as selective as possible, it is well known in biology that cell signaling often proceeds along several parallel pathways, and therefore to obtain a desired pharmacological effect often warrants drugs with multiple "polypharmacological" effects.[11,66] The feature could also be used in concomitant design to reduce activity for an antitarget while increasing it for a target (i.e., increasing the selectivity of a compound). Moreover, one interesting possibility is the design of drugs to overcome drug resistance. As reviewed above, interactions over the mutational space of an HIV drug target are possible. Accordingly, from a theoretical point of view it may be possible to afford inhibitor design over a large area of a drug target's mutational space to achieve better treatment. Proteochemometrics may even allow us to predict new mutations in a target and thus make it possible to design for such future potential mutants even before they arise. It would surely be important to explore this exciting opportunity to find remedies to combat dreadful diseases such as HIV where drug resistance is the major problem in its treatment.

To make all this possible on a large scale, there are some bottlenecks to overcome, which to a large extent relates to the mistreatment of data in current science. The problems of today's scientific culture and ideas as to how to remedy them was discussed in a recent publication by Peter Murray-Rust.[67] In short, how to overcome current problems in science is to promote the use of open data, open source, and open standards. Use of the Internet with open access would allow scientists to publish in a distributed fashion and allow everyone immediate access to all information available. To this end, e-science technologies such as semantic data integration (Web 2.0)[68], distributed computing, and other e-science resources are required. Published research would be available in standardized formats annotated semantically based on established ontologies that will permit efficient information mining of sequences, interaction data, and chemical structures. For example, when initiating a new drug development projects, all previously available data relating to the new project can be retrieved with a single Web search conducted over the entire Internet information content, including specialized databases, to yield all available data that can be used in a proteochemometric analysis. Data analysis takes advantage of the available high-performance computational infrastructure, and visualization and interpretation are integrated fundamentally in to the process. The models obtained can easily be deployed (published) as new e-science components. This integrated approach, where all functionality is available from within a single interface (e.g., Bioclipse) (Figure 6), contrasts fundamentally with the loosely coupled, sequential workflow in Figure 3. This leads to increased productivity for scientists and fewer errors, and allows for more extensive models that have better statistical validity and are straightforward to interpret. Together with proteochemometrics, which affords a systematic analysis of the information previously available, these techniques have the potential to cause drug lead identification and optimization to proceed at an exponential rate; the more previous data that is available, the faster the progress in new projects. The scientific community is on a path toward this type of scientific workflow. Development of standards and ontologies are hot topics in today's research, with HUPO[69] and MGED[70] as examples of organizations that define minimum information criteria and exchange formats in proteomics and gene expression experiments. More and more services are expressed as Web services and are easily integrated into the research pipeline.

We thus predict that proteochemometrics will play an important role in pharmaceutical development as well as in biological research, along with other technologies in which chemical and biological data are utilized in an integrated and parallel fashion, but success reaching its full potential will depend on acceptance among scientists and organizations of the idea of openness and the development of standards and exchange formats and service-oriented architectures, besides the pure technical developments of methods such as proteochemometrics per se. A recent thorough account along these lines was given in ref. 76.

Acknowlegments

Financial support was received from the Swedish VR (04-05957) and Uppsala University (Kof 2007).

REFERENCES

1. Imming, P.; Sinning, C.; Meyer, A. Drugs, their targets and the nature and number of drug targets. *Nat. Rev. Drug Discov*. **2006**, 5, 821–834.
2. Ashburn, T. T.; Thor, K. B. Drug repositioning: identifying and developing new uses for existing drugs. *Nat. Rev. Drug Discov*. **2004**, 3, 673–683.
3. Prusis, P.; Mucenience, R.; Andersson, P.; Lundstedt, T.; Wikberg, J. E. S. A process for identifying the active site in a biological target. 1999. G.B. Patent 99273346.8.
4. Lapinsh, M.; Gutcaits, A.; Prusis, P.; Post, C.; Lundstedt, T.; Wikberg, J. E. S. Classification of G-protein coupled receptors by alignment independent extraction of principal chemical properties of primary amino acid sequences. *Protein Sci*. **2002**, 11, 795–805.
5. Wikberg, J. E. S.; Lapinsh, M.; Prusis, P. Proteochemometrics: a tool for modelling the molecular interaction space. In *Chemogenomics in Drug*

Discovery: A Medicinal Chemistry Perspective, Kubinyi, H.; Müller, G., Eds. Wiley-VCH, Weinheim, Germany, 2004, pp. 289–309.

6. Freyhult, E.; Prusis, P.; Lapinsh, M.; Wikberg, J. E. S.; Moulton, V.; Gustafsson, M. G. Unbiased descriptor and parameter selection confirms the potential of proteochemometric modelling. *BMC Bioinf*. **2005**, 6.
7. Prusis, P.; Uhlen, S.; Petrovska, R.; Lapinsh, M.; Wikberg, J. E. S. Prediction of indirect interactions in proteins. *BMC Bioinf*. **2006**, 7, 167.
8. Strömbergsson, H.; Kryshtafovych, A.; Prusis, P.; Fidelis, K.; Wikberg, J. E. S.; Komorowski, J.; Hvidsten, T. R. Generalized modeling of enzyme–ligand interactions using proteochemometrics and local protein substructures. *Proteins* **2006**, 65, 568–579.
9. Mandrika, I.; Prusis, P.; Yahorava, S.; Shikhagie, M.; Wikberg, J. E. S. Proteochemometric modelling of antibody–antigen interactions using SPOT synthesised peptide arrays. *Protein Eng. Des. Sel*. **2007**, 20, 301–307.
10. Kontijevskis, A.; Prusis, P.; Petrovska, R.; Yahorava, S.; Mutulis, F.; Mutule, I.; Komorowski, J.; Wikberg, J. E. S. A look inside HIV resistance through retroviral protease interaction maps. *PloS Comput. Biol*. **2007**, 3.
11. Wikberg, J. E. S.; Mutulis, F. Targeting melanocortin receptors: an approach to treat weight disorders and sexual dysfunction. *Nat. Rev. Drug Discov*. **2008**, 7, 307–323.
12. Strömbergsson, H.; Daniluk, P.; Kryshtafovych, A.; Fidelis, K.; Wikberg, J. E. S.; Kleywegt, G. J.; Hvidsten, T. R. Interaction model based on local protein substructures generalizes to the entire structural enzyme–ligand space. *J. Chem. Inf. Model*. **2008**, 48, 2278-2288.
13. Lapinsh, M.; Prusis, P.; Mutule, I.; Mutulis, F.; Wikberg, J. E. S. QSAR and proteochemometric analysis of the interaction of a series of organic compounds with melanocortin receptor subtypes. *J. Med. Chem*. **2003**, 46, 2572–2579.
14. Wold, S.; Sjöström, M.; Carlson, R.; Lundstedt, T.; Hellberg, S.; Skagerberg, B.; Wikström, C.; Öhman, J. Multivariate design. *Anal. Chim. Acta* **1986**, 191, 17–32.
15. Cramer, R. D. R.; Patterson, D. E.; Bunce, J. D. Recent advances in comparative molecular field analysis (CoMFA). *Prog. Clin. Biol. Res*. **1989**, 291, 161–165.
16. Klebe, G.; Abraham, U.; Mietzner, T. Molecular similarity indices in a comparative analysis (CoMSIA) of drug molecules to correlate and predict their biological activity. *J. Med. Chem*. **1994**, 37, 4130–4146.
17. Goodford, P. J. A computational procedure for determining energetically favorable binding sites on biologically important macromolecules. *J. Med. Chem*. **1985**, 28, 849–857.
18. Pastor, M.; Cruciani, G.; McLay, I.; Pickett, S.; Clementi, S. Grid-independent descriptors (GRIND): a novel class of alignment-independent threedimensional molecular descriptors. *J. Med. Chem*. **2000**, 43, 323–324.

19. Fontaine, F.; Pastor, M.; Sanz, F.; Incorporating molecular shape into the alignment-free grid-independent descriptors. *J. Med. Chem.* **2004**, 47, 2805–2815.
20. Lapinsh, M.; Prusis, P.; Uhlén, S.; Wikberg, J. E. S. Improved approach for proteochemometrics modeling: application to organic compound–amine G protein-coupled receptor interactions. *Bioinformatics* **2005**, 21, 4289–4296.
21. Vedani, A.; Dobler, M. 5D-QSAR: the key for simulating induced fit? *J. Med. Chem.* **2002**, 45, 2139–2149.
22. Vedani, A.; Dobler, M.; Lill, M. A. Combining protein modeling and 6D-QSAR: simulating the binding of structurally diverse ligands to the estrogen receptor. *J. Med. Chem.* **2005**, 48, 3700–3703.
23. Hellberg, S.; Sjöström, M.; Skagerberg, B.; Wold, S. Peptide quantitative structure–activity relationships, a multivariate approach. *J. Med. Chem.* **1987**, 30, 1126–1135.
24. Sandberg, M.; Eriksson, L.; Jonsson, J.; Sjöström, M.; Wold, S. New chemical descriptors relevant for the design of biologically active peptides: a multivariate characterization of 87 amino acids. *J. Med. Chem.* **1998**, 41, 2481–2491.
25. Gottfries, J. The drug designer's guide to selectivity. *J. Chemometr. Intell. Lab.* **2006**, 83, 148–156.
26. Muthas, D.; Lek, P. M.; Nurbo, J.; Karlén, A.; Lundstedt, T. Focused hierarchical design of peptide libraries—follow the lead. *J. Chemometr.* **2007**, 21, 486–495.
27. Wold, S.; Jonsson, J.; Sjöström, M.; Sandberg, M.; Rännar, S. DNA and peptide sequences and chemical processes multivariately modelled by principal component analysis and partial least-squares projections to latent structures. *Anal. Chim. Acta* **1993**, 277, 239–253.
28. Benson, D. C. Fourier methods for biosequence analysis. *Nucleic Acids Res.* **1990**, 18, 6305–6310.
29. Chou, K. C. Prediction of protein subcellular locations by incorporating quasi-sequence-order effect. *Biochem. Biophys. Res. Commun.* **2000**, 278, 477–483.
30. Dubchak, I.; Muchnik, I.; Holbrook, S. R.; Kim, S. H. Prediction of protein folding class using global description of amino acid sequence. *Proc. Natl. Acad. Sci. USA* **1995**, 92, 8700–8704.
31. Lapinsh, M.; Prusis, P.; Gutcaits, A.; Lundstedt, T.; Wikberg, J. E. S. Development of proteochemometrics: a novel technology of use for analysis of drug–receptor interactions. *Biochem. Biophys. Acta* **2001**, 1525, 180–190.
32. Lapinsh, M.; Veiksina, S.; Uhlén, S.; Petrovska, R.; Mutule, I.; Mutulis, F.; Yahorava, S.; Prusis, P.; Wikberg, J. E. S. Proteochemometric mapping of the interaction of organic compounds with melanocortin receptor subtypes. *Mol. Pharmacol.* **2005**, 67, 50–59.

33. Lapinsh, M.; Prusis, P.; Petrovska, R.; Uhlén, S.; Mutule, I.; Veiksina, S.; Wikberg, J. E. S. Proteochemometric modeling reveals the interaction site for Trp9 modified α-MSH peptides in melanocortin receptors. *Proteins: Struct. Funct. Bioinf*. **2007**, 67, 653–660.

34. Kontijevskis, A.; Petrovska, R.; Mutule, I.; Uhlen, S.; Komorowski, J.; P., P.; Wikberg, J. E. S. Proteochemometric analysis of small cyclic peptides' interaction with wild-type and chimeric melanocortin receptors. *Proteins* **2007**, 83–96.

35. Lapinsh, M.; Prusis, P.; Wikberg, J. E. S. Proteochemometrics modeling of the interaction of amine G-protein coupled receptors with a diverse set of ligands. *Mol. Pharmacol*. **2002**, 61, 1465–1475.

36. Hvidsten, T. R.; Kryshtafovych, A.; Komorowski, J.; Fidelis, K. A novel approach to fold recognition using sequence-derived properties from sets of structurally similar local fragments of proteins. *Bioinformatics* **2003**, 19 (Suppl. 2), ii81–ii91.

37. Lindström, A.; Pettersson, F.; Almqvist, F.; Berglund, A.; Kihlberg, J.; Linusson, A. Hierarchical PLS modeling for predicting the binding of a comprehensive set of structurally diverse protein–ligand complexes. *J. Chem. Inf. Model*. **2006**, 46, 1154–1167.

38. Zhang, S.; Golbraikh, A.; Tropsha, A. Development of quantitative structure-binding affinity relationship models based on novel geometrical chemical descriptors of the protein–ligand interfaces. *J. Med. Chem*. **2006**, 49, 2713–2724.

39. Lapins, M.; Eklund, M.; Spjuth, O.; Prusis, P.; Wikberg, J. E. S. Proteochemometric modeling of HIV protease susceptibility. *BMC Bioinf*. **2008**, 9, 181.

40. HIV Drug Research Centre. http://hivdrc.org. Accessed Nov. 2008.

41. Kontijevskis, A.; Komorowski, J.; J. E. S.; W. Generalized proteochemometric model of multiple cytochrome P450 enzymes and their inhibitors. *J. Chem. Inf. Model*. **2008**, 48, 1840–1850.

42. Prusis, P.; Lapins, M.; Yahorava, S.; Petrovska, R.; Niyomrattanakit, P.; Katzenmeier, G.; Wikberg, J. E. S. Proteochemometrics analysis of substrate interactions with dengue virus NS3 proteases. *Bioorg. Med. Chem*. **2008**, 16, 9369–9377.

43. Chen, J. R.; Chang, B. H.; Allen, J. E.; Stiffler, M. A.; MacBeath, G. Predicting PDZ domain–peptide interactions from primary sequences. *Nat. Biotechnol*. **2008**, 26, 1041–1045.

44. Bock, J. R.; Gough, D. A. Virtual screen for ligands of orphan G protein-coupled receptors. *J. Chem. Inf. Model*. **2005**, 45, 1402–1414.

45. Bock, J. R.; Gough, D. A. Predicting protein–protein interactions from primary structure. *Bioinformatics* **2001**, 17, 455–460.

46. Bock, J. R.; Gough, D. A. Whole-proteome interaction mining. *Bioinformatics* **2003**, 19, 125–134.

47. Chemical Computing Group. http://www.chemcomp.com. Accessed Nov. 2008.
48. Tripos. http://tripos.com. Accessed Nov. 2008.
49. Strand Life Sciences. http://www.strandls.com. Accessed Nov. 2008.
50. Accelrys Software. http://accelrys.com. Accessed Nov. 2008.
51. Talete. http://www.talete.mi.it. Accessed Nov. 2008.
52. MIA S.r.l. http://miasrl.com/. Accessed Nov. 2008.
53. Umetrics. http://www.umetrics.com/. Accessed Nov. 2008.
54. Camo. http://www.camo.com. Accessed Nov. 2008.
55. SPSS. http://spss.com/. Accessed Nov. 2008.
56. SAS. http://www.sas.com/. Accessed Nov. 2008.
57. Clamp, M.; Cuff, J.; Searle, S. M.; Barton, G. J. The Jalview Java alignment editor. *Bioinformatics* **2004**, 20, 426–427.
58. JalView. http://www.jalview.org. Accessed Nov. 2008.
59. The R Project for Statistical Computing. http://www.r-project.org. Accessed Nov. 2008.
60. Spjuth, O.; Helmus, T.; Willighagen, E. L.; Kuhn, S.; Eklund, E.; Wagener, J.; Murray-Rust, P.; Steinbeck, C.; Wikberg, J. E. S. Bioclipse: an open source workbench for chemo- and bioinformatics. *BMC Bioinf*. **2007**, 8, 59.
61. Bioclipse. http://www.bioclipse.net. Accessed Nov. 2008.
62. Neerincx, P. B.; Leunissen, J. A. Evolution of Web services in bioinformatics. *Brief Bioinf*. **2005**, 6, 178–188.
63. Curcin, V.; Ghanem, M.; Guo, Y. Web services in the life sciences. *Drug Discov. Today* **2005**, 10, 865–871.
64. Eclipse. http://www.eclipse.org. Accessed Nov. 2008.
65. Genetta Soft. http://www.genettasoft.com. Accessed Nov. 2008.
66. Zimmermann, G. R.; Lehár, J.; Keith, C. T. Multi-target therapeutics: when the whole is greater than the sum of the parts. *Drug Discov. Today*. 2007, 12, 34–42.
67. Murray-Rust, P. Chemistry for everyone. *Nature* **2008**, 451, 648–651.
68. Web 2.0. http://en.wikipedia.org/wiki/Web2. Accessed Nov. 2008.
69. HUPO. http://www.hupo.org/. Accessed Nov. 2008.
70. MGED Society. http://www.mged.org/. Accessed Nov. 2008.
71. NIMH Psychoactive Drug Screening Program. http://pdsp.med.unc.edu. Accessed Nov. 2008.
72. EMBL-EBI, Swiss-Prot. http://www.ebi.ac.uk/swissprot/. Accessed Nov. 2008.
73. Baldwin, J. M.; Schertler, G. F.; Unger, V. M. An alpha-carbon template for the transmembrane helices in the rhodopsin family of G-protein-coupled receptors. *J. Mol. Biol*. **1997**, 272, 144–164.

74. Perl. http://www.perl.org/. Accessed Nov. 2008.
75. Symyx MDL. http://www.mdli.com/. Accessed Nov. 2008.
76. Wikberg, J. E. S.; Eklund, M.; Willighagen, E. L.; Spjuth, O.; Lapins, M.; Engkrist, O.; Aluarsson, J.; Introduction to Pharmaceutical bioinformatics, 2011 bakleaf, Stockholm ISBN 978-91-979403-0-6.

CHAPTER 4

COMPOUND ACTIVITIES IN TIMES OF SYSTEMS BIOLOGY

DAVID E. PATTERSON
Vistamont Consultancy, Berkeley, California

1 INTRODUCTION

"Systems biology is the study of an organism, viewed as an integrated and interacting network of genes, proteins and biochemical reactions which give rise to life." This definition, taken from the Institute for Systems Biology website (http://www.systemsbiology.org/Intro_to_ISB_and_Systems_Biology/Systems_Biology_the_21st_Century_Science), highlights the basis for seeking to apply systems biology concepts to drug design. A human patient is just such an organism.

A crude computational systems biology view of the connection between a chemical structure and a clinical effect of a drug based on the structure might be thought of as a function:

$$\text{clinical effect} = \mathbf{M}(\mathbf{P}(\mathbf{T}(\mathbf{D}(\mathbf{A}(\{\mathbf{S}\}))))) \tag{1}$$

where $\{\mathbf{S}\}$ is a set of chemical structures administered at a specific dosage; $\mathbf{A}(\cdot)$ is a function reflecting ADME, which distributes the chemical to bodily tissues, modifying the structures along the way in some cases; $\mathbf{D}(\cdot)$ is a function that converts structures to computational descriptors; $\mathbf{T}(\cdot)$ is a function indicating the effect on all molecular targets; $\mathbf{P}(\cdot)$ is the net effect on phenotypic aspects in the organism, such as gene and protein expression changes, after the target effects propagate through the interacting pathways; and $\mathbf{M}(\cdot)$ relates the phenomic response to its full medical effects.

Computational Approaches in Cheminformatics and Bioinformatics, First Edition.
Edited by Rajarshi Guha and Andreas Bender.

Drug discovery is a challenge, because testing of candidate chemicals in the intended human patient is impossible until very late in the game. The pharmaceutical industry has selected a number of surrogate systems in which to evaluate potential new medicinal compounds. Animal models were the primary, if not the only choice of surrogate in the early phases of the industry. By the 1980s, the systems used to evaluate compounds tended increasingly toward relatively simple target-specific biochemical assays, which allowed for testing large numbers of potential candidates rapidly and inexpensively. The measurement of effects of candidate compounds on single molecular targets remains the most widely used surrogate system for early drug discovery and reflects an often unstated notion of "one target, one disease." There are many advantages to such a surrogate, including ease of generation of additional data and straightforward definition of a molecular mechanism of action. Roughly speaking, this surrogate focuses only on values for a single target in the $\mathbf{T}(\cdot)$ function above, removing $\mathbf{A}(\cdot)$.

In such an approach, the goal often is defined as the design of a chemical candidate that has the lowest possible IC_{50} value, that is, the concentration at which a drug has half of its fullest effect. Meeting this goal presumes that the compound has the intended biochemical/target activity at a low enough dosage to avoid most, if not all, off-target effects. Both a wide search for suitable chemicals to screen for activity and the subsequent optimization of preliminary "hit" chemicals for increased activity have been addressed by reducing the chemicals systematically to a set of molecular descriptors, $\mathbf{D}(\cdot)$, applying statistical analyses to relate these descriptors to the available target activity information $\mathbf{T}(\cdot)$ and using the models to select additional chemicals for synthesis and testing that are predicted to have higher activity. The study of a quantitative structure–activity relationship (QSAR) via this process has often proven to be a valuable element in drug discovery.

The early work of Hansch and colleagues[1] proved that hydrophobicity of compounds was key to their activity, and further efforts of the group continue to refine the approach.[2] Since this time, a substantial literature has grown up around successful application of QSAR methods to predict single-target activities of compounds; see, for example, ref. 3 or 4 for a short, yet solid review. Although a number of statistical techniques have been employed, most effort has been directed toward creating better molecular descriptors.

Similar QSAR models can often be built when the surrogate system is a cellular assay rather than a simple biochemical reaction assay. In this case, even if the assay result is still expressed in terms of an IC_{50} value, this value may reflect additional systems biology elements, such as passage of the chemical through the cell membrane, active transport within the cell, metabolism of the chemical, and so on, as well as some of the complex interacting effects of the biological pathways that maintain a living cell—it more nearly reflects the $\mathbf{P}(\cdot)$ level of surrogate system. Thus, the resulting models will probably be less reflective of a specific molecular target binding, as it includes the effects of bioavailability and the net effect of the complexity of the cell, and is more relevant to effects in a patient. As the number and relevance of readouts from the assay increases, the cellular assay reflects more and more closely the underlying systems biology of the cell, if not the organism.

Increasingly, it is recognized that many marketed drugs hit additional molecular targets aside from the target intended.[5–7] This may result in a better medicinal result or may appear as undesirable side effects. In either case, there is evidently a need to evaluate a compound versus a large number of possible targets rather than a single target in the $\mathbf{T}(\cdot)$ formulation.

The focus on specific molecular targets is essential when the design tools are receptor-based methods such as computational docking of a ligand into a protein receptor site or an ion channel, but is not required inherently for ligand-based design tools such as (many) QSAR methods. With ligand-based methods, *any* objective function can be attempted, including one described by patterns of gene or protein expression in a test system for which the pattern is not tied explicitly to known molecular targets. Attempts to derive explicit QSAR models for patterns of phenomic expression are rare if not nonexistent in the literature. An initial study along these lines, first presented by Patterson et al.,[8] is summarized later in the chapter.

The ideal surrogate test system would be a computational patient population that modeled the entire system response for a vast number of patients with varied genetic makeup, environmental history, disease states, and prescription suite, with simulated clinical responses, including off-target effects. This does not yet exist in general form; until then, surrogates such as genomics and proteomics in appropriately complex systems offer a new field for drug design.

2 CURRENT EMERGENCE OF DESIGNED MULTIPLE LIGANDS

The predominant "one target, one disease" paradigm for drug discovery has been changing over the past two decades. Morphy and Rancovic[9] cite the existence of over 300 refereed journal articles dealing with *designed multiple ligands* (i.e., compounds aimed at two or more molecular targets).

These compounds fall into several classes: for example, those that are essentially two separate drugs joined by a cleavable linker (released as separate drugs by metabolism in the body) or by an uncleavable linker, and those that have a partially shared pharmacophore. An unfortunate example of the latter case is the pharmacophore of the hERG potassium channel,[10] which is in many cases a subset of the pharmacophore for GPCR targets, thus making it difficult to design out all risk of cardiac toxicity from hERG. Useful multiple ligands are found in the literature particularly in areas including 5-HT depression compounds, dopamine D2 schizophrenia compounds, histamine H1 allergy drugs, ACE inhibitors for hypertension, NSAIDS for pain relief, and PPAR inhibitors for metabolic diseases.[9]

The capability to reach multiple targets with a single drug, rather than separate drugs, can greatly reduce the complexity of ADME and drug interaction problems [i.e., the $\mathbf{A}(\cdot)$ function of equation (1)]. It is also preferred to reduce the risk of drug resistance, particularly when the molecular targets lie on different pathways in the pathogenic organism. Many existing drugs actually show polypharmacology, despite being designed to be exquisitely selective for a single target. It has been noted that there are relatively few data addressing the question of how current drugs actually work, particularly in terms of which of the 324 known targets are directly affected.[11] Certain areas of research explicitly call for ligands that hit multiple targets, such as anticancer kinase inhibitors (although Gleevec and others were designed to be selective, it has been found that their success is due in part to polypharmacology) and HIV drugs aimed at avoiding drug resistance.[12]

Distinct strategies for designing multiple ligands: Select a starting point with good activity at one target and design in additional activities, or find a starting point with some activity in multiple assays, and optimize from this point. Computational as well as screening results can help direct the optimization; see, for example, ref. 13 for application of QSAR models in this context.

Deciding which targets are desired in the biological profile of a compound is itself nontrivial. If there are something like 324 known drug targets, there are over 50,000 pairs of targets and millions of triplets of targets that one could consider. Even given a partial knowledge of the interacting operative biochemical pathways involved in a disease state, the number of potentially important target combinations can be intimidating. Methods such Monte Carlo simulated annealing[14] have been applied to model biological networks to guide selection of optimized multiple targets when a sufficiently complete description of the underlying relevant pathways is available.

Hopkins[15] notes that network biology (i.e., network analysis of interacting biological pathways) can be used to locate key nodes and links that are most critical to attack for disease treatment. This approach is less likely to result in a drug candidate that, while modifying a molecular target successfully, fails clinically due to the robustness of living systems even to single gene deletions. Careful selection can also reduce the likelihood of lethal side effects from modifying key nodes.

There are multiple ways to approach computer-aided compound design for ligands aimed at multiple targets, $\mathbf{T}(\cdot)$. The conceptually simplest approach takes each target independently to generate a QSAR model, then works with differences among the models to evaluate proposed compounds and to guide the design of new ones. When there is overlap among the targets' pharmacophores, so that target bindings are not independent, it may be better to build models based on explicit recognition of shared factors. The partial least-squares (PLS) approach, for example, identifies latent variables among the molecular descriptors that best explain the latent variables among the full set of target values $\mathbf{T}(\cdot)$.[16] With such results, the mathematical directions to go to best modify chemical parameters to achieve specific biological changes are clearer.

A more recent methodology seeks explicitly to select among potential compounds based on Pareto optimal sets, that is, by finding compounds that are "good enough" in terms of all targets and for which no other compound is better against all targets.[17,18]

3 COMPOUND EXPRESSION SPECTRA AS DESCRIPTORS

As noted, the objective of a drug is to modify a phenomic state in a patient rather than to modify the action of a specific molecular target. Although human patients are not available until very late in the process of drug discovery, there are opportunities to observe the action of a compound in human cellular systems with broad sets of readouts, such as expression-level changes of gene or protein factors. The Broad Institute at MIT has created the Connectivity Map,[19] a resource that evaluates the gene expression spectra of human cells exposed to bioactive compounds at 10 μM. They have successfully demonstrated examples in which new targets for existing drugs were suggested by similarities in their expression spectra, such as valproic acid acting as an HDAC inhibitor.

BioSeek's BioMAP system[20,21] generates data on the expression of a large number of specific proteins in a number of separate human cell systems, which has also been shown to establish useful biological similarity

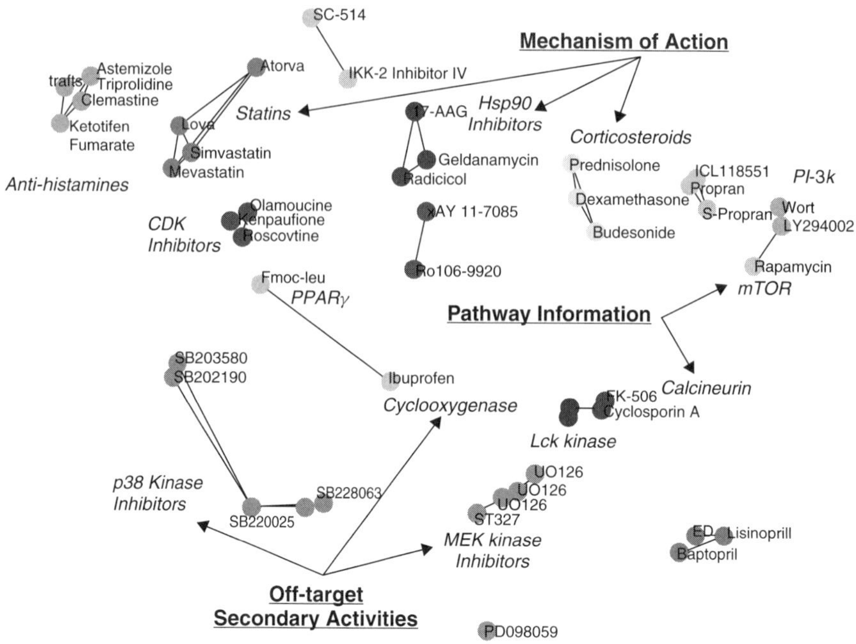

FIGURE 1 Nonlinear map derived from Pearson similarities among readout profiles in BioSeek's BioMAP. Compound profiles which share a target or a pathway or a mechanism of action tend to cluster.

that reflects the mechanism of action as well as pathway information (see Figure 1). Pragmatically speaking, such expression spectra are arguably as close as current technology allows to the full form of equation (1) other than the translation to medical effect. Such a systems biology style assay is particularly desirable for drugs aimed at novel targets with novel chemistry, since there will be great uncertainty about both the exact set of targets being hit and the true systems response to modifying the targets. It is useful to examine a phenomic readout to spot troublesome areas of off-target effects or even lack of intended target effect downstream in a pathway. For drugs in the "me too" category, the imperative is to show improved performance over existing medications, again making it useful to examine details of the system response that most nearly approximates that of a human patient.

4 QSAR APPLIED TO SYSTEMS BIOLOGY EXPRESSION SPECTRA

The genesis of this section is a study conducted by Patterson[22] for BioSeek, Inc., of South San Francisco, California during 2006–2007.

BioMAP Systems are primary cell-based models of human disease biology, designed to replicate the intricate cell and pathway interactions as they are observed in vascular inflammation, cardiovascular and respiratory diseases, fibrosis, and related clinical indications. BioSeek had previously generated BioMAP profiles of differential protein expression in four human cellular systems, 3C, 4H, LPS, and SAg, which included 42 specific readout values measured for a total of 102 compounds within a small-molecule series.[23] With replicates and with several separate concentrations for some compounds, a total of 396 readout patterns were available for the study. The intent was to apply QSAR analysis to understand the biological activity observed, and to apply this understanding prospectively to select additional chemicals that would show similar or improved activity. The molecular target believed to be shared by the compounds was and still is not definitively established, although some phenomenological effects that suggest possibilities have been observed in animal testing for these compounds.

What does one model in this situation? It is not clear that a single molecular target is being affected and there is no information whatsoever about possible binding sites, so target-based SAR methods are inapplicable. It is possible to examine each protein expression readout individually, either quantitatively or as a category of up/down/unregulated. It is preferable to consider the protein expression profile as a whole, but the best parameterization to do so is not immediately apparent.

After investigation of multiple approaches, the Pearson coefficient was finally selected to represent similarity among complete profiles (and was used to generate Figure 1). The results of this prospective application of QSAR in systems biology were nearly but not quite successful, providing a number of insights into the need to improve this initial QSAR modeling process, especially to identify optimal biological descriptors.

Chemical Series

The chemical series remains proprietary to BioSeek, Inc. For the current purposes, the relevant information is illustrated in Figure 2. A central core backbone structure, here represented by the black rectangle, consists of a fused two-ring system containing more than one heteroatom. The training set of compounds contains examples of substitutions at three sites on the backbone, with sample substituents shown here. Substituents varied most among the training set at site 1, with the fewest substituent cases at site 3. There were also small changes within the training set within the backbone, which were not thought to be significant.

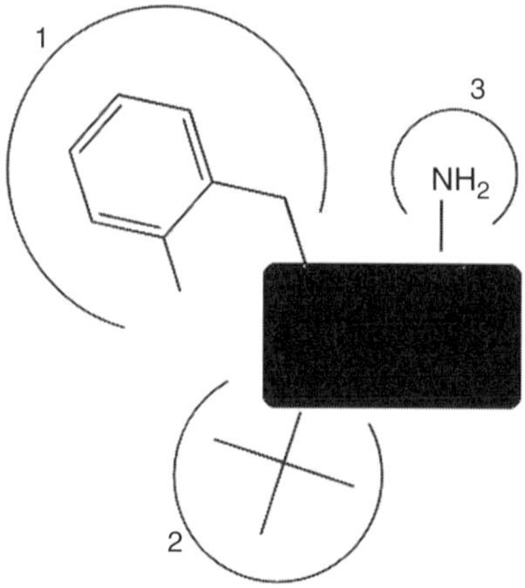

FIGURE 2 Chemical series representation of the training set used in the QSAR in systems biology study.

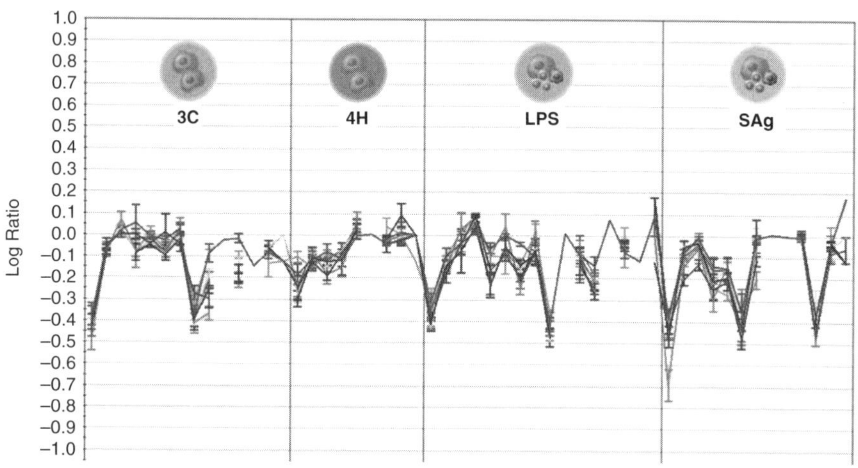

FIGURE 3 Sample replicates from the study as BioMAP readout profiles. Data cover four distinct primary human cell systems and show excellent consistency of the key signals.

Protein Readouts

The set of 42 readouts that represented the systems biology description for each assay were embedded in a set of 52 readouts that included additional measures of cell health and certain morphological measures (Figure 3). The readouts covered four human cellular systems. Each readout represented the logarithm of the ratio of the protein expression level compared to the reference (based on multiple wells per plate for the reference).

For categorical studies, the significance envelope was used to label each such readout as being unchanged (within the envelope), up-regulated, or down-regulated. Some readouts had very few training set examples that were outside the envelope. For this study, no data scaling was performed on a per-row basis. Later work with BioSeek data has suggested that scaling each readout pattern to a unit vector can assist in large-scale classification projects.

QSAR Analysis of Study Data

For all of the present QSAR studies, the training set compounds were aligned based on the conserved central ring scaffold. CoMFA[24] computes descriptors $\mathbf{D}(\cdot)$ as putative interaction energies of a unit charge C atom at each node of a three-dimensional (3D) grid surrounding the molecular collection, such that each molecule has molecular descriptors of steric and electrostatic interaction energy potential at the roughly 1000 grid locations, using a 2-Å grid spacing. These energy grids are then used to generate cross-validated QSAR models. The CoMFA method is among the most widely used 3D QSAR approaches, with hundreds of published applications in the first decade after its appearance.[25]

Direct attempts to use the CoMFA grids, C log P, and compound concentration to explain the levels observed for the 42 readout proteins were not promising, as expected, since none of these readouts were tied directly to a single molecular target. Only 16 of the 42 (38%) had a cross-validated PLS q2 value above 0.50 (which represents an anticipated ability of the model to predict half of the variability of similar new compounds). Identical success rates were found using the related SIMCA discriminant technique to estimate whether each readout would be up- or down-regulated or unchanged.

A PLS-2 analysis was run in which all 42 readouts were treated as one block of dependent variables, seeking common explanations of the readout behavior, and using the same molecular descriptors. *No* predictively useful QSAR model was found for this case. Evidently the phenotype protein expression profile observed is not a simple linear superposition of individual readouts, as network biology would anticipate.

In the process of trying several parameterizations of QSAR modeling, it became clear that the preferred objective was an idealized profile of the entire set of readouts, a profile that existed somewhat imprecisely in the minds of the biologists involved. As a first step in this direction, a best example was selected from the training set and the Pearson correlation coefficient was used as the dependent variable, ranging from 1.0 for the target profile itself to values insignificantly different from zero in the training set. The PLS cross-validated q2 for the model using CoMFA, $C \log P$, and compound concentration to explain the Pearson similarity was marginally successful at 0.62. In this model, steric and electrostatic contributed roughly equally to the result, with minor additive effects from $\log P$ and concentration, and the molecular maps were interpretable.

QSAR generally aims at improving a parameter during lead optimization. In this case, nothing would yield a Pearson above 1.0 (i.e., be more similar to the profile selected than the profile itself). However, the goal was to find similar activity in a different chemical series, since no further synthesis resource was available. Therefore, the CoMFA QSAR model was applied to a set of commercially available compounds that had the same scaffold system—other than atom and bond types—and those available compounds with the highest CoMFA-predicted Pearson values were used to select 22 commercial compounds. Another 15 compounds were selected based on overall steric and electrostatic similarity of structures using the "dbtop" program[26] which has been shown to identify probable activities of compounds based on similarity to compounds of known activity.[8,27,28] Although none of the dbtop selections were as similar to the compound whose profile was used as the objective as was needed to make activity highly likely, moderately similar compounds were chosen, which allowed some further relaxation of the scaffold requirements.

The goal was to find one or more novel compounds that showed experimental Pearson similarity to the readout profile of 0.70 or higher—a value that was judged high enough to avoid large numbers of profiles that were too dissimilar, while not missing many good candidates. In the event, after losing some compounds due to solubility and to cytotoxicity, the test compounds did not find any truly successful experimental profiles. This disappointing result is tempered by several factors, however:

- The objective profile compound when retested along with the new compounds showed only a 0.85 Pearson similarity to its original profile, so that the 0.70 goal may have been ambitious.
- The most nearly successful novel compound selected based on the QSAR prediction, done across chemical series, was 0.70, similar to the target profile compound when tested at a slightly lower concentration, and also to the profile most similar to the target compound among the training set.
- The best dbtop shape similarity selection compound barely missed the 0.70 cutoff, and was also better than 0.70 Pearson to the profile most similar to the target compound among the training set.
- These near misses suggest that there is promise of optimization within the novel compounds; in fact, after relaxing the standards to accept

near concentrations of the target compound and nearest-neighbor compound in the training set, two series within the novel compounds show promise in more than one compound, suggesting that optimization would be worth consideration.

Countering these positive interpretations is the fact that the BioSeek scientists did not find the "near miss" readout profiles to be truly promising. No compound hit all of the most vital readouts that the biologists hoped to see, and it was thus clarified that the unweighted expression spectra Pearson coefficient was not an ideal measure. Perhaps something more like the Tanimoto similarity of up/down/unregulated readout fingerprints as used in the Connectivity Map project would be an improvement. More explicit manipulation of concentration might help, including use of row normalization to selected housekeeping proteins. To some extent, the need is to parameterize concentration in the profile context, as is done via IC_{50} in the single-readout dose–response context. A combination of known biochemical pathway models with empirical deconvolution of the observed readouts into nominal target measures is an ongoing goal for analyzing this sort of systems biology data. The empirical aspect is useful when one does not necessarily know all the potentially relevant targets at work.

The noteworthy aspect of this study is that it contains encouraging signs for future studies applying QSAR models to genomics and proteomic systems. Even with single-target studies it is much simpler to optimize within a chemical series than to leap to entirely new chemical series as was attempted here, and the limited availability of suitable commercially available screening compounds further limited the effort. The lack of complete success in a limited initial study is thus not to be mistaken for a discouraging outcome.

5 FUTURE DIRECTIONS

Extending computer-aided drug design toward matching an ideal gene or protein expression profile shows promise. The field awaits better ways to describe the objective function describing the "omics" profile, better understanding of the systems biology (network biology) relating targets to readouts, and better computational approaches to optimizing a compound toward the objective. Toward the latter point, there is promise in nonlinear control theory[29] methods as applied in complex, incompletely described systems, such as electric power distribution. The mathematics center on questions of observability of the key aspects of the system and on controllability of the system toward a desired state. Realistically speaking, SAR models are simply tools that point the medicinal chemist's thinking in promising directions. Perhaps control systems approaches, such as those under study at the Institute for Systems Biology (see, e.g., ref. 30) will serve a similar function while being better suited to complex, nonlinear system behavior. Ultimately, drug design may be applied[31–34] in a full computational systems biology model of populations to best assess the range of human responses to be anticipated for the new drug candidate.

Acknowledgments

The author wishes to acknowledge the kind support provided by BioSeek Inc. of South San Francisco, California in performing the QSAR study presented here. The guidance and discussions generously offered by Ellen Berg and Eric Kunkel of BioSeek made possible whatever interesting results were obtained.

REFERENCES

1. Hansch, C. A quantitative approach to biochemical structure–activity relationships. *Acc. Chem. Res*. **1969**, 2, 232–239.
2. Hansch, C.; Hoekman, D.; Leo, A.; Weininger, D.; Selassie, C. D. Chembioinformatics: comparative QSAR at the interface between chemistry and biology. *Chem. Rev*. **2002**, 102, 783–812.
3. Debnath, A. K. Quantitative structure–activity relationship (QSAR) paradigm: Hansch era to new millennium. *Mini Rev. Med. Chem*. **2001**, 1, 187–195.
4. Selassie, C. D.; Mekapati, S. B.; Verma, R. P. QSAR: then and now. *Curr. Top. Med. Chem*. **2002**, 12, 1357–1379.
5. Franz, S. Drug discovery: playing dirty. *Nature* **2005**, 437, 942–943.
6. Handl, H. L.; Vagner, J.; Han, H.; Mash, E.; Hruby, V. J.; Gillies, R. J. Hitting multiple targets with multimeric ligands. *Expert Opin. Ther. Targets* **2004**, 8, 565–586.
7. Keiser, M. J.; Roth, B. L.; Armbruster, B. N.; Ernsberger, P.; Irwin, J. J.; Shoichet, B. K. Relating protein pharmacology by ligand chemistry. *Nat. Biotechnol*. **2007**, 25, 197–206.
8. Patterson, D. E.; Cramer, R. D.; Ferguson, A. M.; Clark, R. D.; Weinberger, L. E. Neighborhood behavior: a useful concept for validation of molecular diversity descriptors. *J. Med. Chem*. **1996**, 39, 3049–3059.
9. Morphy, R.; Rankovic, Z. Designed multiple ligands, an emerging drug discovery paradigm. *J. Med. Chem*. **2005**, 48, 6523–6543.

10. Cianchetta, G.; Li, Y.; Kang, J.; Rampe, D.; Fravolini, A., Cruciani, G.; Vaz, R. J. Toward a pharmacophore for drugs inducing the long QT syndrome: insights from a CoMFA study of HERG K(+) channel blockers. *J. Med. Chem.* **2002**, 45, 3844–3853.

11. Overington, J. P.; Al-Lazikani, B.; Hopkins, A. L. How many drug targets are there? *Nat. Rev. Drug Discov.* **2006**, 5, 993–996.

12. Hopkins, A. L.; Mason, J. S.; Overington, J. P. Can we rationally design promiscuous drugs ? *Curr. Opin. Struct. Biol.* **2006**, 16, 127–136.

13. Bender, A.; Young, D. W.; Jenkins, J. L.; Serrano, M.; Mikhailov, D.; Clemons, P. A.; Davies, J. W. Chemogenomic data analysis: prediction of small-molecule targets and the advent of biological fingerprint. *Comb. Chem. High Throughput Screen.* **2007**, 8, 719–731.

14. Yang, K.; Bai, H.; Quyang, Q.; Lai, L.; Tang, C. Finding multiple target optimal intervention in disease-related molecular network. *Mol. Syst. Biol.* **2008**, 4, 228.

15. Hopkins, A. L. Network pharmacology: the next paradigm in drug discovery. *Nat. Chem. Biol.* **2008**, 4, 682–690.

16. Frank, I.; Friedman, J. A statistical view of some chemometrics regression tools. *Technometrics* **1993**, 35, 109–148.

17. Nicolaou, C. A.; Brown, N.; Pattichis, C. S. Molecular optimization using computational multi-objective methods. *Curr. Opin. Drug Discov. Dev.* **2007**, 10, 316–324.

18. Shepphird, J. K.; Clark, R. D. A marriage made in torsional space: using GALAHAD models to drive pharmacophore multiplet searches. *J. Comput.-Aided Mol. Des.* **2006**, 20, 763–771.

19. Lamb, J.; Crawford, E. D.; Peck, D.; Modell, J. W.; Blat, I. C.; Wrobel, M. J.; Lerner, J.; Brunet, J. P.; Subramanian, A.; Ross, K. N.; et al. The connectivity map: using gene expression signatures to connect small molecules, genes and disease. *Science* **2006**, 29, 1929–1935.

20. Butcher, E. C.; Berg, E. L.; Kunkel, E. J. Systems biology in drug discovery. *Nat. Biotechnol.* **2004**, 22, 1253–1259.

21. Kunkel, E. J.; Dea, M.; Ebens, A.; Hytopoulos, E.; Melrose, J.; Nguyen, D.; Ota, K. S.; Plavec, I.; Wang, Y.; Watson, S. R.; Butcher, E. C.; Berg, E. L. An integrative biology approach for analysis of drug action in models of human vascular inflammation. *FASEB J.* **2004**, 18, 1279–1281.

22. Patterson, D. E.; Kunkel, E. J.; Berg, E. L. QSAR Applied in Systems Biology. Oral presentation at the 234th American Chemical Society National Meeting, Boston, Aug. 19–23, 2007.

23. Berg, E. L.; Kunkel, E. J.; Hytopoulos, E.; Plavec, I. Characterization of compound mechanisms and secondary activities by BioMAP analysis. *J. Pharmacol. Toxicol. Methods* **2006**, 53, 67–74.

24. Cramer, R. D.; Patterson, D. E.; Bunce, J. D. Comparative molecular field analysis (CoMFA): 1. Effect of shape on binding of steroids to carrier proteins. *J. Am. Chem. Soc.* **1988**, 110, 5959–5967.

25. Kim, K. H. List of CoMFA References 1993–1997. In *3D QSAR in Drug Design: Recent Advances*, Kubinyi, H.; Folkers, G.; Martin, Y. C., Eds. Kluwer/ESCOM, New York, 1998, pp. 317–338.
26. Cramer, R. D.; Jilek, R. J.; Andrews-Cramer, K. M. Dbtop: topomer similarity searching of conventional structure databases. *J. Mol. Graph. Model*. **2002**, 20, 447–462.
27. Cramer, R. D.; Clark, R. D.; Patterson, D. E.; Ferguson, A. M. Bioisosterism as a molecular diversity descriptor: steric fields of single "topomeric" conformers. *J. Med. Chem*. **1996**, 39, 3060–3069.
28. Cramer, R. D.; Poss, M. A.; Hermsmeier, M. A.; Caulfield, T. J.; Kowala, M. C.; Valentine, M. T. Prospective identification of biologically active structures by topomer shape similarity searching. *J. Med. Chem*. **1999**, 42, 3919–3933.
29. Sontag, E. D. Some new directions in control theory inspired by systems biology. *Syst. Biol*. **2004**, 1, 9–18.
30. Shmulevich, I.; Gluhovsky, I.; Hashimoto, R.; Dougherty, E. R.; Zhang, W. Steady-state analysis of genetic regulatory networks modeled by probabilistic Boolean networks. *Comp. Funct. Genom*. **2003**, 4, 601–608.
31. González-Díaz, H.; González-Díaz, Y.; Santana, L.; Ubeira, F. M.; Uriarte, E. Proteomics, networks and connectivity indices. *Proteomics* **2008**, 8, 750–778.
32. Lill, M. A. Multi-dimensional QSAR in drug discovery. *Drug Discov. Today* **2007**, 12, 1013–1017.
33. Ortiz, A. R.; Gomez-Puertas, P.; Leo-Macias, A.; Lopez-Romero, P.; Lopez-Viñas, E.; Morreale, A.; Murcia, M.; Wang, K. Computational approaches to model ligand selectivity in drug design. *Curr. Top. Med. Chem*. **2006**, 6, 41–55.
34. Vedani, A.; Dobler, M. Multi-dimensional QSAR in drug research: predicting binding affinities, toxicity and pharmacokinetic parameters. *Prog. Drug Res*. **2000**, 55, 105–135.

CHAPTER 5

MOLECULAR DESCRIPTORS FOR BIOLOGICAL SYSTEMS

N. SUKUMAR, SOURAV DAS, MICHAEL KREIN, RAHUL GODAWAT, INNA VITOL, SHEKHAR GARDE, KRISTIN P. BENNETT, and CURT M. BRENEMAN

Rensselaer Exploratory Center for Cheminformatics Research, Rensselaer Polytechnic Institute, Troy, New York

1 WHY USE DESCRIPTORS?

Relating biological activities to calculable properties of molecular structures in a quantitative manner has long been recognized as one of the grand challenges in chemistry and molecular biology. A fervent hope—even an implicit assumption—of most practitioners has been the hypothesis that similar molecules would tend to display similar activities in biological assays. Although often realized in practice, many departures from this guiding principle have been observed in recent years,[1] even with increasing sophistication of machine learning methods designed to extract pertinent information from a proliferation of molecular descriptors. One reason for the departure from this similarity principle is the complex nature of the activity landscape within the existing descriptor space associated with any given biological assay ("activity cliffs"[2–4]). This landscape complexity is inextricably linked with the appropriateness of the chemical space representation (molecular descriptors) and similarity assessment metric employed. Biomolecules have a hierarchy of structures, and thus a hierarchy of associated representations. Often, they possess a wide range of protonation states, depending on the pH, and are generally associated with several molecules of bound water and counterions that critically influence their structure, dynamics and biological functions.

Computational Approaches in Cheminformatics and Bioinformatics, First Edition.
Edited by Rajarshi Guha and Andreas Bender.

The fundamental premise underlying the field of quantitative structure–activity relationship modeling (QSAR)—and indeed, much of molecular biology—is that molecular structure determines biological activity through the physics of intermolecular interactions. However, teasing these relationships apart in the absence of detailed mechanistic understanding of biomolecular processes requires large-scale data collection, statistical modeling or data mining, and a mathematical representation of significant chemical features of molecules. Descriptors are the vehicle by which molecular structure and chemistry are encoded for statistical modeling. A useful set of molecular descriptors needs to encode all molecular characteristics pertinent to observed properties, including interactions with other molecules that result in modulation of specific metabolic pathways. Appropriate representation of these interactions is thus critical to successful application of a QSAR model. Statistical modeling techniques can reveal complex relationships between descriptors and biological activity, but such relationships are typically correlative rather than causative, which also implies that there is no explicit mapping back from a desirable region of activity space to a corresponding domain in molecular structure space. Molecular representation and modeling protocols in bioinformatics and cheminformatics can be considered analogous to languages,[5] with a vocabulary comprised primarily of nucleotide and amino acid sequences (bioinformatics), molecular structures and descriptors derived from them (cheminformatics), and a grammar comprised of a diverse array of searching, machine learning, and data-mining tools.

We begin with a discussion of the use of molecular descriptors in QSAR. In the next section we discuss some challenges faced by molecular descriptors when applied to biological systems, and attempts to surmount these challenges. Several families of descriptors relevant to biological systems are then reviewed in the third section, together with selected applications to modeling problems.

2 CHALLENGES FACED BY MOLECULAR DESCRIPTORS WHEN APPLIED TO BIOLOGICAL SYSTEMS

Although currently, thousands of descriptors are available to the casual user through commercial and academic software packages, clever use of multiple descriptor types typically yields only incremental improvements in model quality. In many schemes this approach only broadens a predictor's applicability domain at the expense of predictive accuracy, indicating that current numerical representations of molecular structures and their associated features might be inadequate. When utilized inappropriately, the addition of descriptors can also serve to narrow the applicability domain of a predictor due to partial overfitting. This provides fertile ground for descriptor research.

The primary task of molecular descriptors for biological systems is to generate a mapping between the relevant biological and chemical spaces, often via descriptions of noncovalent interactions. Relevance and validity remain a concern, since relevance is problem specific, and the primary interaction partners are often not known in advance. When the biological target for a particular problem is identified and its molecular structure solved, structure-based design techniques are the methods of choice. When this information is not known, ligand-based design, virtual high-throughput screens, and QSAR techniques are employed. Several authors[6] favor restricting the descriptor space under consideration to a small set of easily interpretable descriptors. There is much to be said for this strategy: Machine learning models are liable to overfit the training set at the expense of generalizing ability[7,8] as the number of descriptors employed becomes larger. This problem becomes even more acute when working with nonlinear models or with small sets of known data. Models that employ capacity control, such as support vector machines,[9] attempt to minimize the generalization error rather than the training error and can be used to reduce this overfitting tendency. As a result, a compromise has to be made between the interpretability and the predictive ability of a model. In a data-rich environment such as that available from modern high-throughput screens, it is naive to expect a single class of descriptors to capture the complex chemistry of diverse biological phenomena. No single descriptor or set of descriptors is universally applicable. Objective assessment of the applicability domain of models constructed from a given descriptor set is a nontrivial problem that has received considerable attention in recent years.[10–12]

The sizes of biomolecules present a first challenge for descriptor computation, precluding the use of most quantum chemistry–derived descriptors. However, many classes of descriptors can be obtained through classical or semiempirical computations, and a fragment-based approach can fruitfully be employed for others. The hierarchy of structures adopted by proteins and nucleic acids presents a further challenge. Conformational flexibility is a critical link between structure, stability, and biological function. Thus, enzymes need to be flexible enough to mediate a specific reaction pathway, yet rigid enough to achieve molecular recognition. While sequence-based and topological descriptors are insensitive to molecular conformation, three-dimensional (3D) molecular descriptors are often computed from the local minimum-energy conformation, as determined at a particular level of theory. There is, of course, no guarantee that this is the active conformation, or indeed that a single conformation is responsible for the relevant biological activity. Some approaches employ Boltzmann-averaged descriptors to capture the possible contributions of multiple low-energy conformations.[13,14] Because of the complexity of their conformational landscapes, range of protonation, and hydration states, modeling biological systems demands the development and use of descriptors sensitive to conformation, temperature, pH, and hydration. Agrafiotis et al.[15] have reviewed recent developments in molecular mining, structure–activity relationships, chemogenomics, and de novo and fragment-based molecular design and discussed issues related to molecular conformations and solvation.

Descriptors capable of chiral discrimination are yet another challenge. Nine of the top 10 drugs on the market today have chiral active ingredients. For example, although both enantiomers of warfarin are useful as anticoagulants, the $S(-)$-enantiomer of the widely prescribed beta-blocker propanolol is approximately 100 times more potent than the $R(+)$-enantiomer in blocking beta-adrenergic receptors. Similarly, in another chiral beta-blocker, sotalol, the enantiomers have equal class III antiarrhythmic activity, but beta-blocking activity has been attributed mainly to the R-enantiomer. Additionally, the right-handed $S-(+)$-form of the painkiller ibuprofen is three times stronger than the left; L-dopa is used to treat Parkinson's disease, whereas R-dopa is toxic! Recent work has led to the development of novel chiral-selective descriptors that are appropriate for describing biological interactions dependent on chirality.[16,17]

Despite these challenges, structural and sequence descriptors have been employed successfully in bioinformatics because of the phylogenetic relationships between proteins and between genes, leading to a remarkable resilience of protein 3D structures over the course of molecular evolution. These evolutionary relationships can be exploited for structural and functional classification of proteins and genes and for the prediction of protein shapes and functional properties. Protein families clustered on the basis of shape represent strong functional relationships at close range. The biochemical functions of many proteins seem to have been preserved over large evolutionary time scales, and shape similarity often remains detectable even when evolutionary processes may have altered the amino acid sequence beyond recognition.

3 EXAMPLES OF DESCRIPTORS FOR BIOLOGICAL SYSTEMS AND APPLICATIONS TO MODELING PROBLEMS

Descriptors of Classical QSAR

The classical QSAR approach pioneered by Hammett,[18,19] and pursued by Hansch,[20,21] Taft,[22] and others, was based on physicochemical characterizations of molecules. Hammett introduced numerical descriptors to represent substituent groups R on a molecular scaffold in the context of a linear free energy relationship:

$$\frac{\log \mathrm{K}}{\mathrm{K}_0} = \sigma\rho$$

where K is an equilibrium constant and K_0 the equilibrium constant for the reference reaction with $\mathrm{R} = \mathrm{H}, \sigma$ is the substituent constant characteristic of meta or para substitution and depends only on R, and ρ is the reaction constant characteristic of the given reaction. Systematic study with different R groups led to the explanation of both equilibrium and kinetic substituent effects, such as acidity and reaction rates.

The Yukawa–Tsuno equation was a multiparameter extension and reparameterization of the Hammett equation to better account for resonance of the meta- and para-substituted forms:

$$\log \mathrm{K} = \log \mathrm{K}_0 + \rho[\sigma + r(\sigma + -\sigma)]$$

Ortho-substituted substituents introduce steric effects, which are not explained by the equations above. Following an extensive analysis of the reactivities of aliphatic esters, Taft expanded the original equation to account for steric effects:

$$\log \mathrm{K} = \log \mathrm{K}_0 + \rho^*\sigma^* + \delta \mathrm{Es}$$

where σ^* is the polar substituent constant and Es is the steric substituent constant. These relations have been applied extensively over the years to predict pK_a, toxicity, and other physicochemical properties.

Hansch and Leo[21] extended the concept of linear free-energy relationships to describe biological effects of molecules, thereby giving birth to the field of quantitative structure–activity relationships (QSAR). In addition to the Hammett parameter σ that describes electronic effects, Hansch and Leo introduced the lipophilicity parameter π, leading to the QSAR equation:

$$\log\frac{1}{C} = -k\pi + k'\pi^2 + \sigma\rho + k''$$

C is the concentration of a drug needed to achieve a desired biological activity and k, k', ρ, and k'' are regression coefficients that are typically fitted to the training data. Hansch studied the effects of a substituent on the partitioning of molecules between water and 1-octanol, two solvents that model the membrane–aqueous interface in biological systems. The partitioning of a molecule between the two phases is quantified by the equilibrium constant P. The value of the parameter π measures the free-energy change caused by a particular substituent and is obtained as the difference between the substituted and unsubstituted log P values ($\pi = 0$ for hydrogen):

$$\pi_{\mathrm{R}} = \log P_{\mathrm{R}} - \log P_{\mathrm{H}}$$

An extensive review of traditional physicochemical descriptors, linear free-energy relationships, topological indices, spectroscopic descriptors, and comparative molecular field analysis applied to cheminformatics is available in the literature,[23] as is a comprehensive catalog of molecular descriptors employed in cheminformatics.[24]

Descriptors from Biological Assays

Even though the use of experimental data as descriptors precludes prospective prediction of the properties of molecules not yet made, there is much to gain by using one or more known behaviors of an existing molecule to predict other behaviors not yet assayed. For example, as a consequence of the conservation of active site residues within protein families, drug molecules lose the ability to interact selectively with protein targets at a high ligand concentration. Fliri et al.[25,26] have exploited this fact to extract molecular property descriptors from the percent inhibition values determined from a broad spectrum of in vitro assays representing a cross section of the druggable proteome, at a single high drug concentration. A total of 92 assays were employed, encompassing G-protein-coupled receptors, ion channels, proteases, and kinases, and molecules covering a wide range of structural classes were screened, thereby establishing QSARs integrating determinants of both ligands and ligand-binding sites without using information on affinities for specific targets. The bioactivity spectra can be clustered by similarity using any of several metrics, enabling quantitative predictions to be made on the binding of new molecules to multiple proteins.

Balasubramanian et al.[27] used proteomic maps from gel electrophoresis, with each spot in the two-dimensional (2D) gel proteomic map represented by a complex number whose components are charge and mass, to construct atomic adjacency and distance matrices. Eigenvalues and eigenvectors of these matrices then yield novel biodescriptors that characterize proteomics maps and describe the changes induced in cells by various chemicals.

Hawkins et al.[28] have modeled toxicity through the combined use of classical chemical descriptors and biological descriptors. Traditional topological descriptors, path lengths, and information-theoretic indices were augmented with biodescriptors created from information in 2D electrophoresis gels. These described the spot abundances of expressed proteome exposed to 14 halocarbons and a control. It was demonstrated that new information was found in the biodescriptors, and their analysis could lead to additional insight into the mechanism of hepatotoxicity. Zhu et al.[29] also found that the prediction accuracy of k-nearest neighbor QSAR models for chemical toxicity was improved significantly when chemical descriptors were augmented by biological descriptors from HTS data.

Topological Indices and Sequence Descriptors

Kier et al.[30–32] introduced the concept of descriptors as mathematical characterizations of molecular structure. Descriptors commonly employed today range from simple one-dimensional (1D) descriptors based on atom counts and linear sequences to complex multidimensional fingerprints and descriptors computed from quantum chemistry. 1D descriptors are quick to calculate (either through sums of molecular features or other simple arithmetic constructs), and are therefore amenable to high-throughput scenarios; 1D descriptors that define Lipinski's Rule of Five[33] are used as a first step in drug design to filter out virtual screening leads with poor bioavailability. Representation of connectivity gives rise to 2D descriptors, which include topological and fragment descriptors. Topological descriptors are related to fundamental determinants of structure obtained from the molecular graph; they include atom and bond counts, degree of branching, and number of electrons. Topological descriptors derived from the 2D molecular graph are simple to compute, have a long history, and have proven very useful in QSAR. Well-known examples are the Wiener number W, which counts the lengths of all distances between each pair of atoms in a molecule, and Hosoya's topological index Z, which counts all sets of nonadjacent bonds in a structure and the molecular connectivity index, constructed from the row sums of the adjacency matrix.

Randic[34] has reviewed the development of topological indices following the introduction of the connectivity index.

Numerical descriptor encodings of linear DNA sequences[35] have proven useful for labeling the potential hazards of genotoxic chemicals in inducing mutations and deletions in DNA. A graphical representation is used to define a graph radius (and a corresponding distance measure between graphs) that tracks the number and type of specific single and double nucleotide mutations. Stochastic moments of adjacency matrices have been employed as descriptors of RNA secondary structure.[36] These moments, which encode information on RNA electrostatics and secondary structure folding, have been used for the identification and discrimination of bacterial RNA promoter sequences from control sequences.

Sequence and sequence structure comparison have been used widely to predict protein folds. Amino acid composition is the simplest such descriptor, computed as the frequency of occurrence of the amino acids in a given protein:

$$\mathrm{fr}(i) = \frac{N_i}{N}$$

where $i = 1, 2, \ldots, 20$, N_r is the number of amino acids of type i, and N is the length of the sequence, leading to an embedding of sequences in a 20-dimensional feature space. Similarly employing the frequencies of occurrence of vicinal dipeptides as dipeptide composition-based descriptors,

$$\mathrm{fr}(i,\ j) = \frac{N_{ij}}{N - 1}$$

where $i = 1, 2, \ldots, 20$ and N_{ij} is the number of dipeptides composed of amino acids i and j, leads to a sparse and often uneven embedding in a feature space of 400 dimensions. Singular value decomposition may be used to distribute more uniformly the data in a space of reduced dimensionality. Motif methods that use regular expression patterns derived from an analysis of sequences of known function are ubiquitous in bioinformatics.[37–40]

Property descriptors such as hydrophobicity, van der Waals volume, polarity, polarizability, charge, surface tension, flexibility, steric parameters, secondary structure, solvent accessibility, and relative mutability for each residue have been employed[41–44] to compute three aggregate descriptors for each property: composition (C), transition (T), and distribution (D). C measures the frequency of occurrence of amino acids with a particular property, T is a measure of the frequency of an amino acid with a particular property occurring in conjunction with a neighboring amino acid with a different property, and D measures lengths along the protein chain where specified fractions of the amino acids with a particular property are located: for example, the sequence length within which 25% of all the hydrophobic residues are found in a given protein. The feature vector generated from the aggregate descriptors of all properties is used in the Web-based utility SVM-Prot[43] for fold recognition and functional classification of proteins. A similar descriptor scheme has been used for prediction of multihistocompatibility complex binding peptides[45] and of protein–protein interactions.[46] Hybrid methods using the frequencies of consecutive amino acids and amino acid–derived properties have also been used by Gao et al.[47]

Recently, Ong et al.[48] have evaluated the performance of several sets of sequence-derived structural and physicochemical protein descriptors in SVM prediction of six functional families. The descriptor sets examined included amino acid composition, dipeptide composition, autocorrelations of physicochemical properties, and composition, transition, and distribution of structural and physicochemical properties, as well as combinations thereof. The study found no overwhelmingly superior choice of descriptors, but the combination sets were found to give more consistent predictions.

In addition to using sequence information, several methods have incorporated distance and property information to increase rate of recall in similarity searches. Protein chains have been represented as graphs for identification of recurrent patterns of amino acid distributions.[49] Individual amino acids represented by pseudoatoms (a simplified representation of the side chains) form the nodes of the graph, and the edges between nodes capture the geometric relationship between the nodes as internode distances. A subgraph-isomorphism algorithm is used to detect recurrent patterns of triads in a large database of protein structures.

Catalytic triads are a set of three amino acid residues in an enzyme that function together and are involved directly in catalysis. The most common example consists of the residues serine, aspartate, and histidine, which work together to break peptide bonds in certain proteases.[50,51] The catalytic action depends on the interplay of a nucleophile, a general base, and an acid, although the individual components may vary.[52] The residues of a catalytic triad might be far from each other in the primary structure, but are close together in the tertiary structure and exhibit similar spatial arrangements within protein families. Catalytic triads thus function much like pharmacophores (see below) in small molecules. Patterns in amino acid sequences in binding sites of proteins have also been studied to infer functional behavior.[53] Even when overall sequence identities between primary sequences were low, sequence identities of concatenated short chains of amino acids forming the binding site surface were significantly higher, demonstrating evolutionary conservation.

Topological autocorrelation descriptors are analogous to molecular connectivity indices commonly used in cheminformatics. Autocorrelation functions of sequence-derived properties have also been employed for amino acid sequences characterizing proteins[47,54–56]:

$$A(d) = \sum_{i} P_i P_{i+d}$$

where P_i and P_{i+d} are the values of a particular property for a pair of amino acids i and $i + d$ separated by d residues along the protein sequence. Alternative formulations use deviations of the properties from the average values instead of the property values themselves as the basis for autocorrelation or the squared difference of property values instead of vector products. Autocorrelation descriptors measure the correlation of a property with itself measured along the sequence (topological autocorrelations) or through 3D space (spatial autocorrelations).

Fragment descriptors are representations of local atomic environments. As the description of the local environment becomes more specific, these descriptors, referred to as fingerprints, are used for substructure searching and as input for molecular similarity analysis. Commonly used fingerprint descriptors include Daylight, Unity from Tripos, and MDL MACCS fragment keys from Elsevier, where each hashed portion or "key" describes a substructure of fewer than 10 heavy atoms. Fast searching may be accomplished by storing the absence or presence of these keys as a vector of indices, allowing for rapid comparison of molecules for biologically related activity in virtual high-throughput screens. Molecular holograms, such as those generated in Tripos's *Sybyl*, extend keyed fingerprints by storing frequency of appearance of features and not simply their presence or absence. Fingerprints enable rapid similarity searches of large databases but are not useful for modeling continuous responses. Similarly, floating-point descriptors may be more effective than fingerprints in assessing similarity (or producing kernel functions) in activity space.

Unity fingerprints have been utilized to extract molecular structural features responsible for bitter taste perception,[57] a biological phenomenon mediated by multiple receptors. Nidhi et al.[58] used extended connectivity fingerprints (ECFPs) of small molecules to predict the top three protein targets of the 964 protein targets in the World of Molecular Bioactivity chemogenomics database. The descriptors, developed by SciTegic, are circular substructural fingerprints based on the Morgan algorithm.[59] An ECFP feature represents an exact structure with limited and specified attachment points, incorporating information from nearest neighbors, next-nearest neighbors, and so on, iteratively. Once a neighborhood of specified granularity is reached, the set of all features is returned as a fingerprint, resulting in descriptors that can describe both global and local features. A neighborhood size of six was chosen in the computation of ECFPs used as structural descriptors. Multiple category naive Bayes models were created from the descriptors. Nidhi et al. found that, on average, the correct target was predicted in the top three guesses 77% of the time; 79% of the time, the correct family was predicted. The authors suggested that these descriptors coupled with the models could be used to predict off-target drug effects, and would be amenable to novel drug detection.

Bender et al.[60] used atom environment MOLPRINT 2D descriptors and a naive Bayes classifier to outperform Unity fingerprints in conjunction with fusion of ranking scores and binary kernel discrimination methods in 10 of 11 classes. Sybyl atom types were used in their definition, atom environments being calculated from the molecular connectivity table. The use and success of both ECFP and MOLPRINT 2D descriptors in enhancing recall in similarity searches and target prediction suggest that information captured from molecular neighborhoods boosts selectivity and enables scaffold hopping over traditional 2D descriptors.

Linear string sequences can also be employed as descriptors for small molecules. A familiar example is the encoding of molecular structures by SMILES (simplified molecular input line entry system) strings. Grant et al.[61] employed a decomposition of canonical SMILES strings into overlapping character substrings (*Lingos*), which were then used in QSAR models, thus reducing the task of molecular similarity assessment to comparison of *Lingos*. A computationally efficient implementation of the method, using deterministic finite-state automata, for rapid similarity searching of molecular databases has been demonstrated.

3D Descriptors

Three-dimensional descriptors can better represent geometry-dependent biological activity and typically encode conformational information either by alignment of molecules in an interaction field,[62] by autocorrelation of vectors,[63] or by other means.[64–69]

Pharmacophores Not long after Fischer proposed the "lock and key" mechanism of enzyme-binding substrate, Ehrlich[70] described a pharmacophore as "a molecular framework that carries (*phoros*) the essential features responsible for a drug's (*pharmacon*) biological activity." Pharmacophores are thus special arrangements of features (aromatic, lipophilic, positively charged, negatively charged, hydrogen bond donor/acceptor, etc.) where atoms are defined based on their membership in the aforementioned feature classes. The information content of molecular fingerprints can also be extended by incorporating conformational sensitivity. Three-dimensional molecular fingerprints are commonly based on pharmacophore representations of molecules. In many cases, however, 3D descriptors perform worse than expected (or even than 2D descriptors)[60,71] since if the conformations used to compute descriptors are inappropriate,[72] the additional descriptors contribute primarily noise to the model.

Reducing the number of chemical features can help in rapid 3D comparison of chemical structures from a large database with relatively high reliability. Such a "fuzzy" 3D similarity search–based approach was used in FEPOPS, a system developed at Novartis[73] to find drug leads rapidly through "scaffold hopping." For each conformer of a molecule, four centroids were created by clustering all atoms. Properties such as the sum of partial charges within a cluster, the sum of atomic log P values, the presence of hydrogen bond donors and acceptors, and distances to other centroids were mapped onto these clusters. FEPOPS similarity between two molecules was defined as the best Pearson's correlation coefficient between their aligned FEPOPS conformers.

Molecular Interaction Fields Biological activity depends critically on molecular shape and on interactions that are often noncovalent in nature. Molecular mechanics force fields, which usually treat noncovalent interactions only as steric and electrostatic forces, are sufficient to model a large number of biomolecular properties. The comparative molecular field analysis (CoMFA) technique[62] samples 3D steric and electrostatic fields around sets of ligands aligned to best describe the binding bioactive conformations. The method is aimed at identifying chemical features favoring or disfavoring the biological activity of interest, which are then employed in a three-dimensional QSAR equation (3D-QSAR). A set of molecules with measured activity form the training set. The molecules are aligned in 3D space by superimposition techniques that maximize the steric and electrostatic overlap or by virtue of a pharmacophore model. The predictive quality of a 3D-QSAR model depends largely on the quality of alignment and its resemblance to the actual bioactive conformation. Once a suitable alignment is obtained, charges are then calculated for each molecule at a desired level of theory. The next step is to construct a field. A probe atom or group of atoms is chosen that computes steric and electrostatic fields for each molecule by interaction at a series of grid points surrounding the aligned data set of molecules in space. The values at each grid point are then used to construct a 3D-QSAR equation. CoMFA further employs isocontour plots to represent electrostatically and sterically favorable or unfavorable areas around the molecules, for chemical interpretation of the models.

Cramer et al.[74,75] suggested a protocol for generating invariant 3D representations of molecular fragments from their 2D topologies, through deterministic rules specifying the absolute configuration, conformation, and orientation. These "topomers" have been employed for shape similarity searching of very large virtual libraries and to provide a solution to the CoMFA alignment problem. Topomers are characterized by CoMFA-like steric shapes and by pharmacophoric features. The steric similarity of two topomers is computed as the squared sum of differences in steric field values over corresponding pairs of lattice points. Thus, topomer shapes are compared as a combination of fragment-to-fragment differences, involving a single conformation for each fragment. In contrast to pharmacophore-based 3D searching, where shape comparison focuses on a small set of atomlike features, topomer shape comparison considers all atoms. Cramer et al.[74,75] have benchmarked the performance of topomers versus Tanimoto similarity on 2D fingerprints and demonstrated that topomer similarity searching enabled scaffold hopping among different classes of structures.

Local Molecular Surface Property Descriptors

Clark et al.[76,77] have argued that descriptors based on local molecular surface properties that do not encode the chemical constitution of the molecule directly, are likely to favor scaffold hopping and lead to more global QSAR/QSPR models. Some examples of local surface properties properties are listed in Table 1.

Polar Surface Area and Solvent Accessible Surface Area Descriptors Polar surface area (PSA) descriptors account for the surface areas of polar atoms (N, O) and atomic fragments (N–H, O–H) in the 3D structure of molecules.[78] They have been applied in QSAR models for a wide range of biological properties, such as blood–brain partitioning,[79,80] intestinal absorption,[81–84] and oral bioavailability.[85] PSA descriptors obtained from a single conformation have yielded predictive models similar to those obtained from averaging all probable conformations.[86] Similar approaches involve scaling of surface areas by partial atomic charges[87] and use of partitioned total surface area, quantifying PSA, and total surface area as separate descriptors.[88]

TABLE 1 Surface Electronic Properties

EP	Electrostatic potential	$\text{EP} = \Sigma_\alpha Z/_\alpha \lvert r - R_\alpha \rvert - \int \rho(r')dr'/\lvert r - r' \rvert$
BNP	Bare nuclear potential	$\text{BNP} = \Sigma_\alpha Z/_\alpha \lvert r - R_\alpha \rvert$
MLP1	Audry's molecular lipophilicity potential[174]	$\text{MLP1} = \Sigma_i f_i/(1 + d_i)$
MLP2	Heiden's molecular lipophilicity potential[175]	$\text{MLP2} = \Sigma_i f_i g(d_i)/\Sigma_i g(d_i)$ $g(d_i) = [\exp(-a \cdot d_{\text{cutoff}}) + 1]/[\exp(a(d_i - d_{\text{cutoff}})) + 1]$
$\nabla\rho \cdot \mathbf{n}$	Electron density gradient normal to 0.002 e/bohr^3 electron density isosurface	$\nabla\rho(r) \cdot \mathbf{n}$
G	Electronic gradient kinetic energy density	$G(r) = -\nabla\psi * \cdot \nabla\psi$
K	Electronic Schrödinger kinetic energy density	$\text{K(r)} = -(\psi * \nabla^2\psi + \psi \nabla^2\psi*)$
$\nabla K \cdot \text{n}$	Gradient of the Schrödinger kinetic energy density normal to the surface	$\nabla \text{K(r)} \cdot n$
$\nabla G \cdot n$	Gradient of the gradient kinetic energy density normal to the surface	$\nabla G(r) \cdot n$
F^-	Fukui electrophilic reactivity index[176]	$F^- = [\delta\rho(r)/\delta dN]^-_{v(r)}$
F^+	Fukui nucleophilic reactivity index[176]	$F^+ = [\delta\rho(r)/\delta dN]^+_{v(r)}$
F^0	Fukui radical reactivity index[176]	$F^0 = ([\delta\rho(r)/\delta dN]^-_{v(r)} + [\delta\rho(r)/\delta dN]^+_{v(r)})/2$
L	Laplacian of the electron density[95]	$L(r) = -\nabla^2\rho(r)$
PIP	Local average ionization potential	$\text{PIP} = -\Sigma_{i=1}^{\text{HOMO}} \rho_i\varepsilon_i/\rho(r)$
EA_L	Local average electron affinity	$\text{EA}_L = -\Sigma_{i=\text{LUMO}}^{N} \rho_i\varepsilon_i/\rho(r)$
α_L	Local polarizability[76]	$\alpha_L = \Sigma_i \rho_i q_i \alpha_i/\Sigma_i \rho_i q_i$
η_L	Local hardness	$\eta_L = (\text{PIP} - \text{EA}_L)/2$

Solvation effects are implicated in protein folding and stability, whereas desolvation is associated with protein–protein and protein–ligand binding. The alternative to explicit treatment using water molecules in a simulation[89] is the use of implicit solvent models[90] that do not take into account the solvent degrees of freedom explicitly, but treat the solvent as a continuum having average properties of the actual solvent. The solvent accessible surface area (ASA) model is an example of an implicit solvent model[91] that measures the solvation free energy of a solute as a sum of atomic contributions, weighted by their solvent-exposed area.

A modification of the ASA model, the Coulomb accessible surface area model, includes screening effects of the solvent on electrostatic interactions between atoms of a protein.[90,92]

Guha and Jurs[6] modeled a data set consisting of 211 artemisinin and artemisinin analogs, which are potent antimalarial compounds. Since chirality-insensitive descriptors were used, the authors removed enantiomers with lower potency, resulting in a total of 179 molecules. A series of descriptors that form part of the ADAPT methodology were used. This included geometric (moments of inertia, molecular surface areas, and volumes), topological (path lengths, connectivity indices, and electrotopological state), electronic (HOMO and LUMO energies, electronegativity, and hydrogen bonding), and hybrid descriptors (charged partial surface area descriptors and hydrophobicity descriptors). From this set of descriptors, a process of objective feature selection removed highly correlated descriptors and descriptors with low variance. Linear (partial least-squares-based) and nonlinear (neural-network-based) models were built from the final set of descriptors selected. The final models were built on a subset of the reduced set of descriptors chosen by a subjective feature selection using genetic algorithm and neural networks. The final nonlinear model had better predictive ability; however, the authors pointed out that the slightly less predictive linear model could easily be interpreted in terms of the structural requirements (encoded by descriptors) for activity of artemisinin analogs.

Quantum Chemical Descriptors Quantum chemical descriptors derived from molecular electron density distributions are applicable to a wide variety of small molecules, but the intensive computational effort required to generate them through ab initio calculations precludes their routine use for large biomolecules. This limitation is circumvented in the RECON algorithm[93,94] by employing an atoms-in-molecules-based approach[95,96] for the rapid computation of molecular electronic properties from atomic charge density fragments precomputed from ab initio wavefunctions. Atomic fragments that satisfy Bader's virial partitioning prescription[95] have uniquely defined properties that are approximately additive and transferable from one molecule to another. The molecular descriptors are constructed in most cases by simple arithmetic operations on the respective atomic descriptors stored in a precomputed transferable atom equivalent (TAE) library, making the method well adapted for virtual high-throughput screening applications, scaling well with both molecular and database size (with throughputs on the order of 1 million molecules an hour on a single-processor Linux computer).

Some commonly used electronic TAE descriptors include the gradient of the electron density normal to the molecular surface ($\nabla\rho \cdot \mathbf{n}$), the electrostatic potential (EP),[97,98] and the bare nuclear potential (BNP), mapped onto the molecular van der Waals surface, the rate of change of the electronic kinetic energy density normal to and away from the molecular surface, Politzer's local average ionization potential (PIP),[99,100] the Fukui electrophilic, nucleophilic, and radical reactivity indices (F),[101,102] and the Laplacian distribution or the trace of the second derivative matrix of the electron density (Table 1).

Although simple histogram distributions of these molecular surface properties are insensitive to molecular conformation, a computationally inexpensive implementation (RAD–RECON autocorrelation descriptors) that incorporates 3D shape information within the TAE RECON formalism is achieved through property autocorrelations binned by the distance R_{xy} between atom pairs x and y for each TAE property P:

$$A(R_{xy}) = \frac{1}{n} \sum_{x,y} P_x \; P_y$$

n being the number of bins. An analogous scheme using amino acid fragments has been employed to generate TAE descriptors for proteins[103,104] and polypeptides.[56] Local position-dependent descriptors were employed by Doytchinova et al.[105] to give predictive regression models for the binding affinities of polypeptide sequences to the HLA-A*0201 major histocompability complex. Bergeron et al.[56] found improved predictive ability in nonlinear models for this data set upon incorporation of topological autocorrelation descriptors derived from TAE. This study also employed a new type of similarity (SIMIL) score between pairs of amino acid residues. Each SIMIL score was a two-part score, consisting of a class score and a RECON score. The class score was a weighted score constructed out of bits representing the presence of the following physical characteristics: tiny, small, positive, negative, polar, nonpolar, aliphatic, and aromatic. The RECON score was constructed from weighted differences of TAE RECON descriptors. These SIMIL scores appear as a 20×20 similarity matrix. Amino acid properties such as hydrophobicity, hydrophilicity, side-chain mass, and volume have long been used to compute chemical distance matrices between sequences and sequence order correlation factors.[106,107]

Several methods for prediction of protein–DNA binding specificity have been developed using structural information from protein–DNA complexes. Both the primary sequence and the 3D conformation of a protein seem to contribute to specifying the binding site in DNA-site

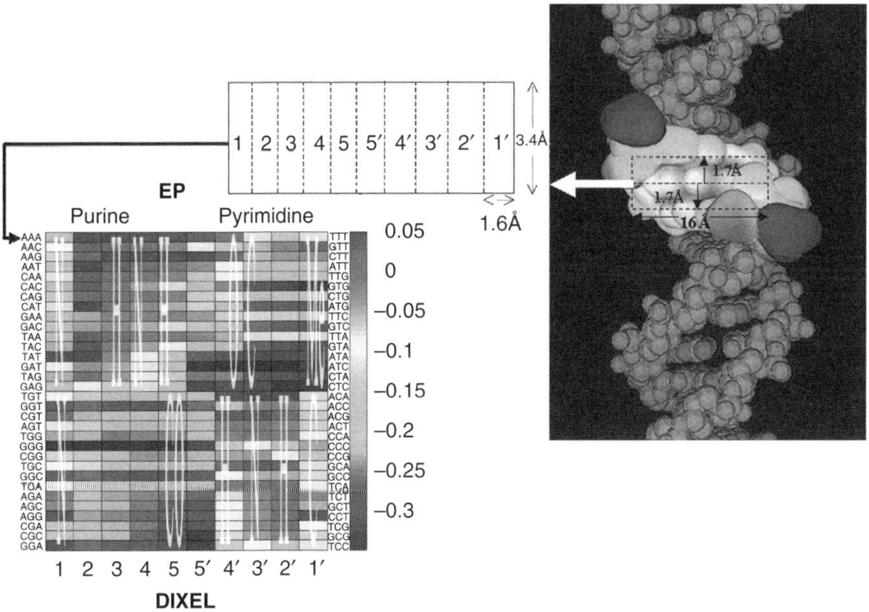

FIGURE 1 Mapping from the DNA major groove surface to a Dixel representation. Each base pair is represented by a rectangular grid of surface pixels. The resulting library of base pair triples is employed to reconstruct the DNA sequence based on the surface electron density properties of the central base pair of each triplet. (*See insert for color representation of the figure*.)

recognition, leading to the use of hybrid models combining structural and primary sequence information. Electron density characterization of DNA has been employed to reconstruct the chemical properties of DNA fragments using a variant of the TAE method. This characterization was accomplished by extracting electron density information from ab initio electronic structure calculations on all possible sets of three stacked base pairs, with the central base pair residing in the specific electronic environment generated by the flanking base pairs. Figure 1 shows a mapping from the DNA major groove surface to a Dixel representation, where each base pair is represented by a rectangular grid of surface pixels. The resulting library of base pair "triples" was then employed to reconstruct the DNA sequence based on the surface electron density properties of the central base pair of each triplet. The local electronic environments induced by neighboring base pairs were observed to have a strong influence on electronic properties, such as EP and PIP, as shown in Figure 2 for several possible base pair triples.

Three broad classes of methods have generally been used for predicting target sites of transcription factors: sequence-base, energy-based, and

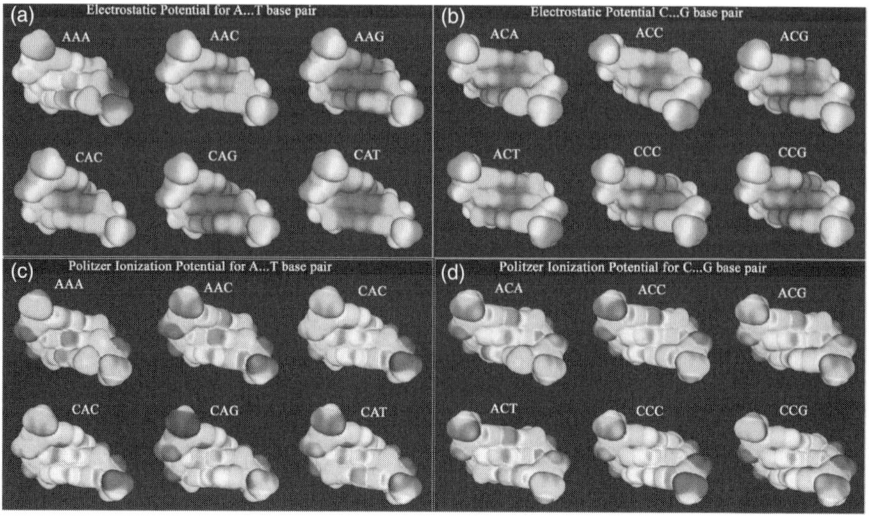

FIGURE 2 Local electronic environments induced by neighboring base pairs on the electronic properties (a, b) EP and (c, d) PIP for several possible base pair triples. The central base pair is an A···T base pair in (a) and (c) and a C···G base pair in (b) and (d).

structure-based.[108] To date, the most successful computational schemes for the identification of these sites are motif methods based on models that represent DNA polymers by sequences of letters that seek to identify the characteristic sequence patterns of short spans of DNA sequence. Numerous algorithms have been developed to identify motifs from multiple observations, including Gibbs sampling,[109,110] greedy consensus algorithms,[111] and expectation maximization (EM) algorithms.[112–115] However, these motif modeling models are limited by an abstract representation of DNA as letters that has little to do with the energetics of the binding of proteins to DNA. This limitation can be addressed using a more fundamental characterization of the DNA polymer, through the use of selected electron density properties encoded on the surfaces of the major and minor groves of DNA (Dixel descriptors). A data set of binding sites for the sigma 70 transcription factor of *Escherichia coli* and a set of control sequences from intergenic regions of convergently transcribed genes and from upstream regions of tandem transcribed genes were studied[116,117] with respect to the ability of the models to discriminate the sets of sequences of 29 nucleotides of *E. coli* DNA most likely to be sigma factor–binding sites from control sequences. Classification based on sequence alone performs quite well on this task: the naive Bayes approach (NB) of creating two generative models under the assumption of independence of the bases, followed by the application of Bayes' rule,[118] managed to eliminate almost all the control sequences. However, among the several thousand control sequences, some were always predicted to be binding sites with high probability. The procedure consisted of incorporating the Dixel data and the sequence representation into a hybrid procedure, with kernel partial least squares regression[119] for the Dixel variables, to compute a function to reduce the residual classification error on the training data. The addition of Dixel variables, EP, BNP and PIP, to sequence variables was able to capture higher-order effects (Figure 3) and to reduce classification error. Since DNA/protein-binding site identification and quantification is a key component of DNA bioinformatics and gene regulation research, the availability of Dixel descriptors to translate DNA sequences into chemically relevant information will provide a data-rich environment for testing bionformatics models.

Alignment-Free Shape Descriptors

ROCS[68] is a fast shape comparison method for virtual screening, based on the idea that molecules have similar shape if their volumes have significant overlap; any volume mismatch is a measure of dissimilarity. ROCS perceives molecular shape[120] as continuous functions constructed from atom-centered Gaussians. The shape of a query molecule is used to "scaffold hop" to molecules that are similar in their three-dimensional shapes, but that may have low similarities to the query molecule in terms of their (2D) bond frameworks.

Although intended originally for use in diversity indices instead of QSAR applications, BCUT descriptors, developed by Pearlman, Burden, and Smith, have proven to be useful in that domain as well.[121–124] BCUTs are nonintuitive descriptors derived from the eigenvalues of a Burden adjacency matrix but have been shown to capture information that is complementary to that provided by more traditional descriptor types.[125]

The Zauhar shape signatures method[67,69] involves a ray-tracing procedure within the interior of the molecular envelope (defined by the van der Waals or solvent-accessible surface) and collecting ray-length and angle-of-reflection information at each point of intersection of the ray with the surface (Figure 4). Shape signatures encode molecular shape through the distribution of ray lengths, thereby rapidly generating distinctive, compact "shape signature" fingerprints for each molecule without the need for performing computationally intensive 3D alignments between molecules. The property encoded surface translator (PEST)[66] method extends the concept of shape signatures by recording TAE surface property information at each point of intersection of the ray with the molecular surface to produce hybrid shape-property descriptors useful for both similarity assessment

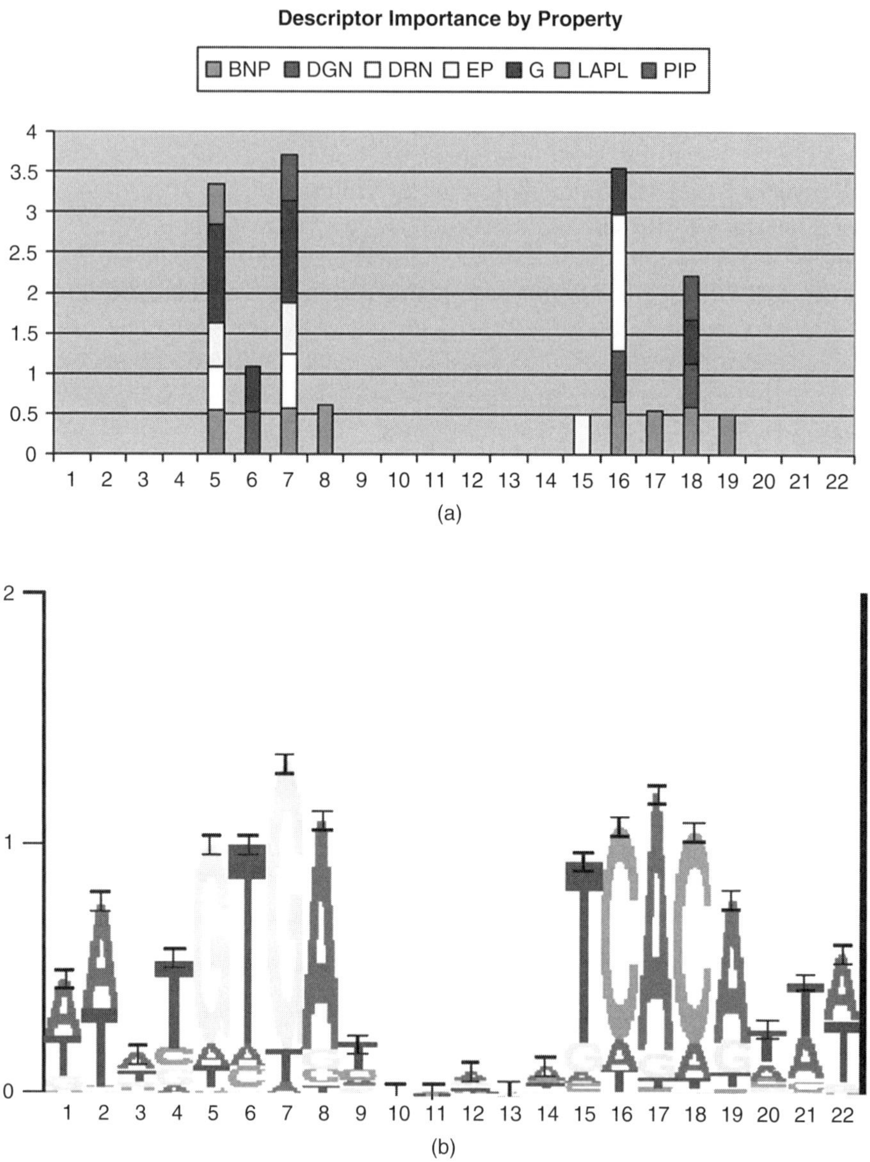

FIGURE 3 (a) Dixel prediction of CRP binding sites, shown along with (b) the experimental consensus footprints.

and QSAR/QSPR machine learning applications. Surface properties that have been employed fruitfully in PEST include electrostatic potentials from either ab initio computation or empirical charges, other TAE electronic properties, and molecular lipophilicity potentials. Descriptors are encoded as 2D histograms (Figure 5) and wavelet coefficients.[126]

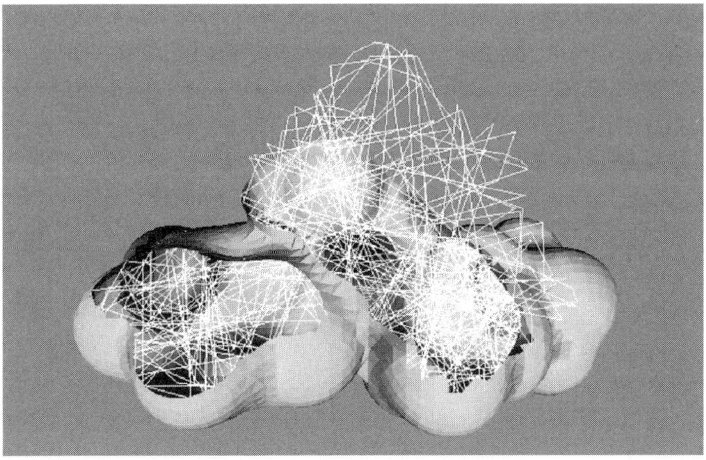

FIGURE 4 Property-encoded ray tracing from a sparse 750-segment PEST run, with part of the molecular van der Waals surface cut away to show the ray bounces. (*See insert for color representation of the figure*.)

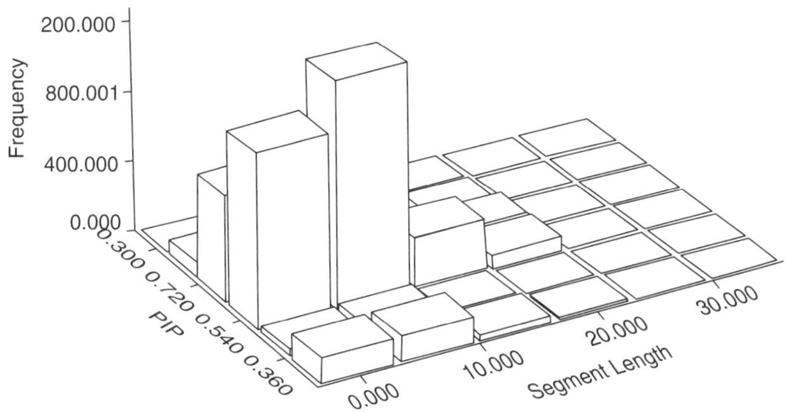

FIGURE 5 EP shape/property two-dimensional histogram for the molecule in Figure 4.

GRIND descriptors by Cruciani et al.[13,127,128] represent another class of alignment-independent shape and property features computed by the Almond package, which are complementary to those available from the PEST or shape signatures approach. GRIND descriptors encode molecular interaction field distributions at key points around a molecule in the form of correlograms rather than capturing molecular shape or surface information. The molecular interaction fields used by GRIND and Almond may be computed using GRID, Volsurf, or other empirical functions.[129]

Ultrafast shape recognition (USR) was developed as a similarity search tool by Ballester and Richards.[64,65] USR descriptors are molecular shape moments with respect to a small set of well-defined points within the molecule: namely, the centroid (*ctd*), the closest atom to the *ctd* (*cst*), the farthest atom from the *ctd* (*fct*), and the farthest atom from the *fct* (*ftf*). Like shape signatures or PEST, USR is alignment-free, generates a compact shape profile, has been shown to perform well at shape classification, and is extremely fast, well suited for high-throughput applications. The RECON algorithm further generates rapid shape-electronic-property hybrid descriptors for high-throughput screening by computing TAE property moments with respect to *ctd, cst, fct*, and *ftf*:

$$\mathrm{USP}_{mk} = \sum_{i} P_i R_{ik}^{m}$$

for each TAE property P, where the summation i runs over all atoms in the molecule, $m = 1, 2, 3$ (corresponding to first, second, and third moments), and $k \in ctd$, *cst, fct, ftf*.

Molecular Quantum Similarity Measures Molecular quantum similarity measures encode electron density–derived information into quantitative measures of molecular similarity.[130–144] The degree of similarity can be established on the basis of the electronic distribution, the topology of the bond critical points[138–140,143,144] or any electron density–derived property. Leherte et al.[145] developed a procedure for similarity searching of molecules based on a comparison of critical point representations of 3D electron density maps and applied it for similarity searching of pharmaceutical databases.

WHIM, 3D MoRSE Code, and GETAWAY Descriptors Todeschini et al.[146–148] have employed principal components treatment of atomic coordinates to design a set of WHIM (weighted holistic invariant molecular) shape descriptors. In this procedure, the atomic coordinates are centered and weighted covariance matrices are computed, which then form the basis of a PCA. WHIM descriptors, invariant to molecular translation and rotation, are extracted from the corresponding scores matrix without the need for molecular alignment. WHIM descriptors can be computed for any set of 3D coordinates and have been used successfully in modeling of physicochemical properties, toxicity, and biological activity. Consonni et al.[149,150] have also constructed new molecular representations and 3D GETAWAY (geometry, topology, and atom weights assembly) descriptors using a leverage matrix, the molecular influence matrix:

$$\mathbf{H} = \mathbf{M} \cdot (\mathbf{M}^{\mathbf{T}} \cdot \mathbf{M})^{-1} \cdot \mathbf{M}^{\mathbf{T}}$$

with elements h_{ij}, where $\mathbf{M}$ is the molecular matrix and $\mathbf{M}^{\mathbf{T}}$ its transpose. The diagonal elements h_{ii} (or leverages) of the molecular influence matrix **H** encode the influence of each atom in determining the overall shape of the molecule. Thus, peripheral atoms have higher leverage than atoms near the molecular center. The maximum leverage in a molecule depends on the molecular size and shape, with lower leverages for atoms in spherical molecules and higher leverages for atoms in linear molecules; the maximum leverage decreases as the molecular size increases. Off-diagonal elements h_{ij} represent the degree of accessibility of atom j to interactions with atom i. The simplest GETAWAY descriptor constructed from the **H** matrix is the geometric mean on the leverage magnitude:

$$H_{\mathrm{GM}} = 100(\Pi_{i=1}^{A} h_{ii})^{1/A}$$

where A is the number of atoms in the molecule. The H_{GM} index increases with branching and decreases with molecular size. The **H** matrix can also be used to define information indices, local vertex invariants, and autocorrelation descriptors. Another matrix, the influence/distance matrix, can be constructed with off-diagonal elements:

$$\mathbf{R}_{ij} = \frac{(h_{ii} h_{jj})^{0.5}}{r_{ij}}$$

with r_{ij} the geometric distance between atoms i and j with leverages h_{ii} and h_{jj}. The **R** matrix can be used similarly to define connectivity indices, branching indices, and autocorrelation descriptors. Consonni et al.[149,150] have applied the H- and R-GETAWAY descriptors in several QSAR and QSPR studies and benchmarked their performance with respect to other constitutional, topological, and shape descriptors, finding a combination of WHIM and GETAWAY descriptors to be useful for predictive modeling of properties dependent on the 3D features of molecules.

Gasteiger et al.[151] employed a molecular transform used in electron diffraction studies to devise a representation of 3D molecular structure that is characterized by a fixed number of variables, independent of the size of the molecule:

$$I(s) = \sum_{i=2}^{N} \sum_{j=1}^{i-1} P_i P_j \frac{\sin (s r_{ij})}{s r_{ij}}$$

Values of $I(s)$ were calculated from the 3D molecular structure for 32 evenly distributed values of $s = 0, \ldots, 31\ \text{Å}^{-1}$ and constitute the 3D MoRSE (molecule representation of structures based on electron diffraction) code of the property P. Properties employed included mass, atomic number, charge, polarizability, and electronegativity, as well as combinations of different properties.

Protein Shape Comparisons Alignment of molecular structures for shape comparison has a long tradition in both cheminformatics and bioinformatics, based on the structure–property similarity principle. The 3D structure of a residue sequence and the way it folds in 3D space are the main factors determining the biological functions of a protein. Since primary sequence largely determines protein structure, shape comparison for proteins can be represented in three dimensions as a rigid-body superimposition; in two dimensions as similar patterns in distance matrices; or as an alignment of amino acid sequences in one dimension.[152] The DALI Web application[153] performs automated database similarity searches of new solved protein structures against structures in the protein databank (PDB), employing 3D coordinates to compute residue–residue distance matrices and thence an optimal pairwise alignment of protein structures and hierarchical classification into folds. Despite low sequence similarity, high structural similarity is an indicator of a possible phylogenetic relationship. Multiple alignments of distant related proteins can reveal invariant conserved residues that cluster together in 3D to form an active site even if they may be widely separated along the 1D sequence.

Another method for producing rotation-invariant descriptors of the 3D shapes of proteins[154] employs polar Fourier transforms on triangulated surfaces to produce rotation-invariant descriptors of the 3D shapes of proteins. Morris et al.[155,156], using the method of Ritchie and Kemp,[157] developed an effcient method for comparing protein-binding pockets based on spherical harmonics expansion. Spherical harmonics Y_{lm} form a complete set of orthonormal functions, indexed by two integers, l and m. The expansion coeffcients, c_{lm}, are obtained by spectral decomposition. Due to orthonormality, the expansion coeffcients are unique and can therefore be used directly as a feature vector for describing the shape. The lower l values correspond to low frequencies and describe the rough low-resolution shape, whereas the higher values add high-resolution detail to the picture. Truncation of the expansion at a particular value of l corresponds to a low-pass filter. The advantages of using spherical harmonics are that the method is analytical with relatively few parameters and results in unique expansion coeffcients for a given shape. Complex-valued spherical harmonics can be combined to give real functions that are better suited for describing real surface distributions while sharing the same orthonormality and completeness properties. Spherical t-designs and precomputed nearly equal weight integration layouts[158] were used to reduce the computational effort significantly in the calculation of spherical harmonic expansion coefficients. Shapes were compared using a standard Euclidean metric, and an orientation convention using moments of inertia was employed to obviate the need to search for optimal alignment. Spherical harmonic expansion thus offers a compact and fast shape description and comparison method for protein-binding pockets, although it is not well suited for finding local shape features or when the ligand is only partially bound in the binding site. The idea of using spherical harmonic expansion as structural descriptors is akin to that of wavelet decomposition.[126]

3D Motif Models In contrast to global shape comparison methods such as DALI, 3D motif methods look for local structural motifs. Nebel et al.[159] generated 3D motifs based on consensus atom positions without explicit reference to residues. Ligand-binding sites of proteins were compared by superimposing the corresponding rigid ligands. A similarity metric between ligand environments was generated from the number of common atoms sharing a spatial neighborhood:

$$S_{AB} = 0.5\left(\frac{C_{AB}}{T_A} + \frac{C_{AB}}{T_B}\right)$$

where C_{AB} is the number of atoms within the pairwise ligand binding site and T_A and T_B are the number of atoms in the ligand-binding sites of proteins A and B, respectively. This permits clustering of ligand-binding sites into consensus 3D motifs associated with specific biochemical functions.

Protein structures are commonly characterized in terms of the Ramachandran map, which reveals correlations between the backbone dihedral angles (φ and ψ). Different regions of the Ramachandran map correspond to qualitatively distinct structural motifs, such as the α-helix, the β-sheet, and the left-handed helix. However, higher-order correlations have also been discovered,[160] extending across dihedral angles spanning multiple residues, which are useful descriptors of protein structure. Mezei[161] introduced novel fingerprint descriptors, defined without the use of distances, for characterization of protein folds. These fingerprints take the form of binary matrices, whose elements are defined by the angles between the C=O direction, the backbone axis, and the line connecting the α-carbons of the amino acid residues. Rogen and Bohr[162] employed knot invariants to construct a family of global geometric measures for protein structure characterization and classification. These measures were shown to carry more information than secondary structure content, and further, enabled proteins of different sizes to be compared directly without need for alignment.

Protein Protonation and Hydration State PPEST (the protein property-encoded surface translator)[163] was developed for describing the shape and surface property distributions of proteins. It uses a technique akin to ray-tracing to explore the volume enclosed by a protein. Probability distributions were derived from the ray trace, based on the geometry of the reflecting ray and include joint dependence on properties such as the molecular lipophilicity potential and molecular electrostatic potential (EP). These probability distributions, stored as histograms or wavelet coefficients, create a unique orientation independent profile for each protein. The triangulated protein surface subjected to internal ray reflection is derived from the water accessible surface generated by a mesh surface algorithm.[164,165] The profiles generated by PPEST (Figure 6) can be compared rapidly to test for similarity between proteins.

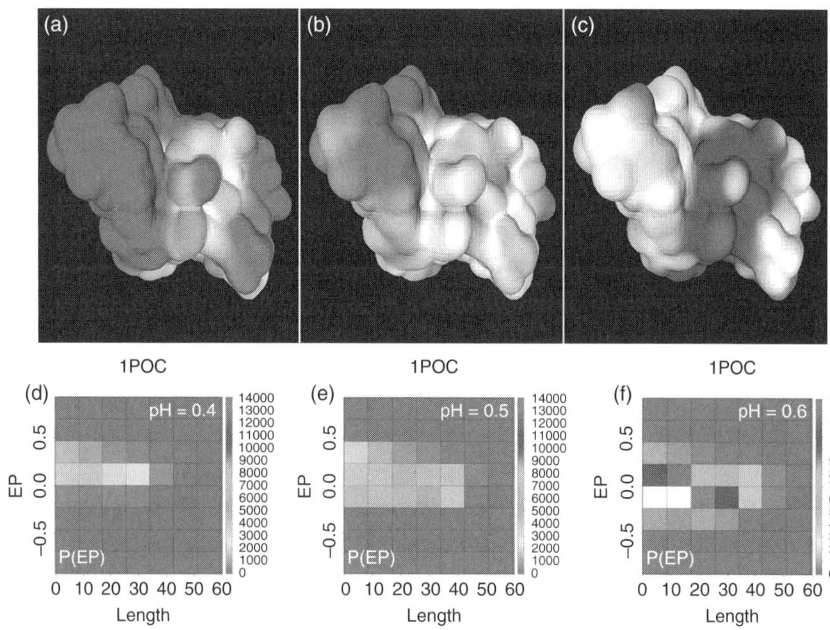

FIGURE 6 EP surface profiles from PPEST for the protein 1POC at three different values of pH [(a) pH 4.0, (b) pH 6.0, and (c) pH 8.0], and corresponding shape/property histograms [at (d) pH 4.0, (e) pH 6.0, and (f) pH 8.0]. (*See insert for color representation of the figure*.)

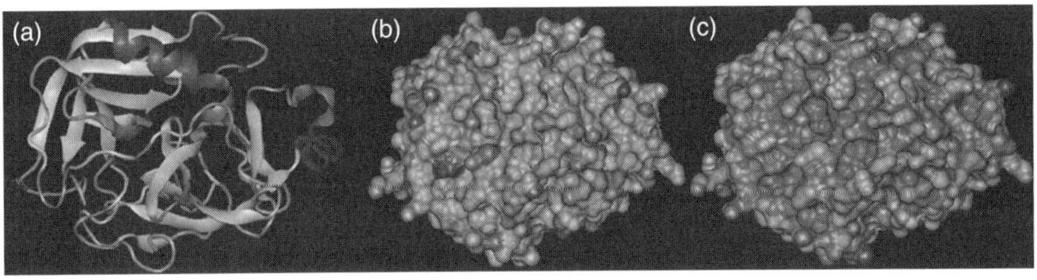

FIGURE 7 (a) Traditional backbone ribbon structure, (b) electron density distribution, and (c) water oxygen fluctuations from MD simulations for the CspB protein.

Interactions of proteins and other biomolecules with water critically influence the properties of vicinal water molecules, an effect that is especially pronounced locally, in the first hydration shell. Statistical analysis of the dynamic behavior of water distributions solvating large and small molecules was used to compute a set of regional property descriptors, such as the average local water density, water density fluctuations, local water orientations, EP on a protein surface induced by the vicinal water structuring, and dynamics of local water, were projected on to a triangulated surface of the molecule, for use as surface descriptors (Figure 7). These descriptors were found to generate better ligand-binding models than traditional field- or atom-based descriptors.[166]

Garde et al.[167] have used a potentials-of-mean-force (PMF) expansion approach, employing a precomputed library of hydration two- and three-particle correlation functions between protein constituent sites and water molecules, derived from explicit molecular dynamics simulations. This was employed to predict the average equilibrium densities and orientation profiles of water in the spaces surrounding biomolecules and ligands. The method approximates the effective free energy of interaction of a biomolecule with surrounding water in terms of two-particle and multiparticle PMF between constituent sites and water molecules. Truncation of the expansion at the three-particle level provided sufficient accuracy for application to hydration of libraries of several hundreds or thousands of molecules of biological or pharmacological interest.

Delaunay Tessellation–Based Descriptors TAE descriptors have recently been combined with atom-type-based fuzzy distance binning for prediction of the pK_d and pK_i of protein–ligand complexes.[168] The protein surface atoms in contact with the ligand were identified by Delaunay tessellation methods and distance cutoffs and their TAE descriptors computed and summed. Atoms within a zone of 10 Å to the ligand were "fuzzy binned" with respect to their atom types and distances from other atoms in the zone. Fuzzy binning is a procedure that increases the count of the neighboring bins by some magnitude in addition to increasing the value of the intended bin by the same or greater magnitude. The binned atom-type distances were then combined with the TAE descriptors and trained over a set of known pK_d and pK_i values to generate the final scoring function.

The complementary ligands based on receptor information (CoLiBRI) approach[169] uses TAE descriptors to search for ligands complementary to binding sites. The method is based on the representation of both receptor binding sites and their respective ligands in a space of universal chemical descriptors, calculated independently for each ligand and for its active site atoms. The binding site atoms involved in the interaction with ligands were identified by Delaunay tessellation applied to x-ray-characterized ligand–receptor complexes. Using the same set of chemical descriptors for ligands and for active sites enabled correlation of chemical similarities between active sites and their respective ligands to be elucidated, using a protocol for mapping patterns of nearest-neighbor active site vectors in a multidimensional TAE space onto those of their complementary ligands, and vice versa. This enabled prediction of a virtual complementary ligand vector in the ligand chemical space from the position of a known active site vector, and identification of molecules in the chemical database most similar to the virtual ligand. Conversely, possible complementary receptor cavities could be identified starting from the ligand chemical structure. Oloff et al.[169] showed, using a data set of 800 x-ray-characterized ligand–receptor complexes in the PDBbind database, that knowledge of the receptor active site structure enabled identification of the known complementary ligand among the top 1% of a large chemical database in over 90% of all test cases when a binding site of the same protein family was present in the training set. When the test receptors were highly dissimilar and not present among the receptor families in the training set, CoLiBRI was still able to quickly eliminate 75% of the chemical database as improbable ligands.

4 SUMMARY AND CONCLUSIONS

While an exhaustive treatment of all molecular descriptors in current use[24] is beyond the scope of this chapter, we have described several classes of descriptors appropriate for biological molecules, ranging from sequence-based, physicochemical or biological assay-based and topological (graph theoretical) to global and local 3D interaction fields, surface properties, and shape signatures and discussed some of the challenges of descriptor-based modeling methods in applications to biological systems. Descriptor utility can be characterized by both interpretability and predictive ability on molecules not included in the training set. These are often mutually conflicting goals, and the best compromise depends on which of these two objectives is more crucial for a given problem. A proliferation of descriptors in a model can often lead to overfitting,[8] the "curse of dimensionality," and a tendency for the model to "memorize" the peculiarities of the training data, which leads to poor generalizing ability. This tendency is more acute in the case of more complex, nonlinear models. Statistical modeling techniques employing capacity control seek to minimize the generalization error (defined as the sum of the training error and the model complexity) rather than the training error, and can yield more robust models with better predictive ability on molecules not included in the training set. Chance correlations can also be caught by sensitivity analysis, where all descriptors having less correlation than an introduced random variable with the activity to be modeled are dropped, and the Y-scrambling test,[11] where the activities to be modeled are permuted among the molecules, and any model predicting these permuted activities equally well or almost as well as the real responses is considered suspect.

Despite the increasing prevalence of high information content and novel descriptors of both biological and chemical importance, methods to characterize the usefulness of such descriptors have hitherto lagged behind, but are now increasingly occupying the attention of researchers. Gone are the days when a good cross-validation was sufficient for a model to be considered a "good model." [170] The focus of QSAR is now not only on predictive ability but also on recognition of the domain of applicability of models[10–12]; that is, when does it make sense to apply a model within a problem domain, and where is that problem domain with respect to the model? Attempts to answer these questions have included characterization of the information content of descriptors,[171] the usefulness of QSAR models in terms of the number of samples available for training,[172] and the stability of models characterized by Shannon entropy, Kendall's tau, and the Spearman correlation coefficient.[173] Guha and Van Drie[3,4] have addressed the identification and the quantification of activity cliffs in chemical models of biological activity by defining a structure–activity landscape index. In representing pairwise sets of molecules as a graph, Guha and Van Drie were able to highlight abrupt changes in response, suitably identifying and characterizing the activity cliffs of biological activities. These steps pave the way for descriptor-agnostic assessments of model quality and model applicability. The use of inappropriate descriptors and injudicious use of high-capacity modeling methods without appropriate validation are responsible for most of the problems reported in QSAR/QSPR modeling.

Statistical modeling and pattern mining technologies and the development of common, standard ontologies for cheminformatics and bioinformatics can enable a rich cross-fertilization of computational methods developed independently in different disciplines. Descriptors are the vehicle by which the vocabulary of sequence information, molecular structure, and chemistry is encoded for this arsenal of statistical modeling tools. The future of prospective structure–property modeling of biological systems depends in large part on the development and availability of descriptors capable of representing the underlying physics of causative intermolecular interactions. Appropriate representation of sequence and molecular structure through well-designed descriptors rich in information content, coupled with objective assessment of model applicability, are crucial to solving the grand challenge of quantitatively relating biological function to macromolecular chemistry.

REFERENCES

1. Martin, Y. *J. Med. Chem*. **2002**, 45, 4350–4358.
2. Maggiora, G. M. *J. Chem. Inf. Model*. **2006**, 46(4), 1535.
3. Guha, R.; Van Drie, J. H. *J. Chem. Inf. Model*. **2008**, 48(8), 1716–1728.
4. Guha, R.; Van Drie, J. H. *J. Chem. Inf. Model*. **2008**, 48, 646–658.
5. Sukumar, N.; Krein, M.; Breneman, C. M. *Curr. Opin. Drug Discov. Dev*. **2008**, 11(3), 311–319.
6. Guha, R.; Jurs, P. *J. Chem. Inf. Comput. Sci*. **2004**, 44, 1440–1449.
7. Cronin, M. T. D.; Schultz, T. W. **2003**, 622(1–2), 39.
8. Hawkins, D. M. *J. Chem. Inf. Comput. Sci*. **2004**, 44, 1–12.
9. Vapnik, V. N. *Statistical Learning Theory*, Wiley, Chichester, UK, 1998.
10. Nikolova, N.; Jaworska, J. *QSAR Comb. Sci*. **2003**, 22, 1006–1026.
11. Eriksson, L.; Jaworska, J.; Worth, A. P.; Cronin, M. T. D.; McDowell, R. M.; Gramatica, P. *Environ. Health Perspect*. **2003**, 111(10), 1361–1375.
12. Breneman, C. M. Model applicability domains: When can I use my model? Presented at the 235th ACS National Meeting, New Orleans, LA, 2008.
13. Cruciani, G.; Crivori, P.; Carrupt, P. A.; Testa, B. **2000**, 503(1–2), 17.
14. Krarup, L. H.; Christensen, I. T.; Hovgaard, L.; Frokjaer, S. *Pharm. Res*. **1998**, 15(7), 972.
15. Agrafiotis, D. K.; Bandyopadhyay, D.; Wegner, J. K.; van Vlijmen, H. *J. Chem. Inf. Model*. **2007**, 47(4), 1279–1293.
16. Golbraikh, A.; Tropsha, A. *J. Chem. Inf. Comput. Sci*. **2003**, 43(1), 144.
17. Marrero Ponce, Y.; Diaz, H. G.; Zaldivar, V. R.; Torrens, F.; Castro, E. A. *Bioorg. & Med. Chem.* **2004**, 12(20), 5331–5342.

18. Hammett, L. P. *Chem. Rev*. **1935**, 17, 125.

19. Hammett, L. P. *J. Am. Chem. Soc*. **1937**, 59, 96.

20. Hansch, C.; Muir, R. M.; Fujita, T.; Maloney, P. P.; Geiger, F.; Streich, M. *J. Am. Chem. Soc*. **1963**, 85, 2817–2824.

21. Hansch, C.; Leo, A. *Exploring QSAR*. ACS, Washington, DC, 1995.

22. Taft, R. W. *J. Am. Chem. Soc*. **1952**, 74, 3120.

23. Livingstone, D. J. *J. Chem. Inf. Comput. Sci*. **2000**, 40, 195–209.

24. Todeschini, R.; Consonni, V. *Handbook of Molecular Descriptors*. Wiley-VCH, New York, 2000.

25. Fliri, A. F.; Loging, W. T.; Thadeio, P. F.; Volkmann, R. A. *J. Med. Chem*. **2005**, 48, 6918–6925.

26. Fliri, A. F.; Loging, W. T.; Thadeio, P. F.; Volkmann, R. A. *Proc. Natl. Acad. Sci. USA* **2005**, 102(2), 261–266.

27. Balasubramanian, K.; Khokhani, K.; Basak, S. C. *J. Proteome Res*. **2006**, 5, 1133–1142.

28. Hawkins, D. M.; Basak, S. C.; Kraker, J.; Geiss, K. T.; Witzmann, F. A. *J. Chem. Inf. Model*. **2006**, 46(1), 9–16.

29. Zhu, H.; Rusyn, I.; Richard, A.; Tropsha, A. *Environ. Health Perspect*. **2008**, 116(4), 506–513.

30. Kier, L. B.; Hall, L. H. *Molecular Connectivity in Chemistry and Drug Research*. Academic Press, New York, 1976.

31. Kier, L. B.; Hall, L. H. *Molecular Connectivity in Structure–Activity Analysis*. Research Studies Press, Letchworth, UK, 1986.

32. Kier, L. B.; Murray, W. J.; Randic, M.; Hall, L. H. *J. Pharmacol. Sci*. **1975**, 64, 1971–1974.

33. Lipinski, C. A.; Lombardo, F.; Dominy, B. W.; Feeney, P. J. *Adv. Drug Discov. Rev*. **2001**, 46(1–3), 3.

34. Randic, M. *J. Mol. Graph. Model*. **2001**, 20(1), 19–35.

35. Nandy, A.; Basak, S. C. *J. Chem. Inf. Comput. Sci*. **2000**, 40(4), 915–919.

36. González-Díaz, H.; Pérez-Bello, A.; Uriarte, E. *Polym. Blends Compos. Hybrid Polym. Mater*. **2005**, 46(17), 6461.

37. Sigrist, C. J. A.; Cerutti, L.; Hulo, N.; Gattiker, A.; Falquet, L.; Pagni, M.; Bairoch, A.; Bucher, P. *Brief Bioinf*. **2002**, 3, 265–274.

38. Suzuki, M. *Structure* **1994**, 2, 317–326.

39. Suzuki, M.; Yagi, N. *Proc. Natl. Acad. Sci. USA* **1994**, 91, 12357–12361.

40. Mandel-Gutfreund, Y.; Margalit, H. *Nucleic Acids Res*. **1998**, 26(10), 2306–2312.

41. Dubchak, I.; Muchnik, I.; Mayor, C.; Dralyuk, I.; Kim, S. H. *Proteins* **1999**, 35, 401–407.

42. Dubchak, I.; Muchnik, I.; Holbrook, S. R.; Kim, S. H. *Proc. Nat. Acad. Sci. USA* **1995**, 92(19), 8700–8704.

43. Cai, C. Z.; Han, L. Y.; Ji, Z. L.; Chen, X.; Chen, Y. Z. *Nucleic. Acids Res.* **2003**, 31(13), 3692–3697.
44. Cai, Y. D.; Liu, X. J.; Xu, X. B.; Chou, K. C. *J. Comput. Chem.* **2002**, 23, 267–274.
45. Cui, J.; Han, L. Y.; Lin, H. H.; Zhang, H. L.; Tang, Z. Q.; Zheng, C. J.; Cao, Z. W.; Chen, Y. Z. *Mol. Immunol.* **2007**, 44(5), 866.
46. Lo, S. L.; Cai, C. Z.; Chen, Y. Z.; Chung, M. C. *Proteomics* **2005**, 5, 876–884.
47. Gao, Q.-B.; Wang, Z.-Z.; Yan, C.; Du, Y.-H. *FEBS Lett.* **2005**, 579(16), 3444.
48. Ong, S.; Lin, H.; Chen, Y.; Li, Z.; Cao, Z. *BMC Bioinf.* **2007**, 8(1), 300.
49. Artymiuk, P. J.; Poirrette, A. R.; Grindley, H. M.; Rice, D. W.; Willett, P. *J. Mol. Biol.* **1994**, 243(2), 327.
50. Carter, P.; Wells, J. A. *Nature* **1988**, 332, 564–568.
51. Dodson, G.; Wlodawer, A. *Trends Biochem. Sci.* **1998**, 23(9), 347.
52. Polgar, L. *Cell. Mol. Life Sci.* **2005**, 62, 2161–2172.
53. Binkowski, T. A.; Adamian, L.; Liang, J. *J. Mol. Biol.* **2003**, 332(2), 505.
54. Lin, Z.; Pan, X. M. *J. Protein Chem.* **2001**, 20, 217–220.
55. Chou, K. C.; Cai, Y. D. *J. Chem. Inf. Model.* **2005**, 45(2), 407–413.
56. Bergeron, C.; Hepburn, T.; Sundling, M.; Sukumar, N.; Katt, W. P.; Bennett, K. P.; Breneman, C. "Prediction of peptide bonding affinity: Kernel methods for nonlinear modeling" http://arxiv.org/abs/1108.5397.
57. Rodgers, S.; Busch, J.; Peters, H.; Christ-Hazelhof, E. *Chem. Senses* **2005**, 30, 547–557.
58. Nidhi, M. G.; Davies, J. W.; Jenkins, J. L. *J. Chem. Inf. Model.* **2006**, 46(3), 1124–1133.
59. Morgan, H. L. *J. Chem. Doc.* **1965**, 5, 107–113.
60. Bender, A.; Mussa, H. Y.; Glen, R. C.; Reiling, S. *J. Chem. Inf. Comput. Sci.* **2004**, 44(1), 170–178.
61. Grant, J. A.; Haigh, J. A.; Pickup, B. T.; Nicholls, A.; Sayle, R. A. *J. Chem. Inf. Model.* **2006**, 46, 1912–1918.
62. Cramer, I.; Richard, D.; Patterson, D. E.; Bunce, J. D. *J. Am. Chem. Soc.* **1988**, 110, 5959.
63. Wagener, M.; Sadowski, J.; Gasteiger, J. *J. Am. Chem. Soc.* **1995**, 117, 7769–7775.
64. Ballester, P. J.; Richards, W. G. *Proc. R. Soc. A* **2007**, 463, 1307–1321.
65. Ballester, P. J.; Richards, W. G. *J. Comput. Chem.* **2007**, 28, 1711–1723.
66. Breneman, C. M.; Sundling, C. M.; Sukumar, N.; Shen, L.; Katt, W. P.; Embrechts, M. J. *J. Comput.-Aided Mol. Des.* **2003**, 17, 231–240.
67. Nagarajan, K.; Zauhar, R.; Welsh, W. J. *J. Chem. Inf. Model.* **2005**, 45, 49–57.

68. Rush, T. S.; Grant, J. A.; Mosyak, L.; Nicholls, A. *J. Med. Chem*. **2005**, 48(5), 1489–1495.
69. Zauhar, R.; Moyna, G.; Tian, L.; Li, Z.; Welsh, W. J. *J. Med. Chem*. **2003**, 46, 5674–5690.
70. Ehrlich, P. *Dtsch. Chem. Ges*. **1909**, 42, 17.
71. Oprea, T. *J. Braz. Chem. Soc*. **2002**, 13(6), 811–815.
72. Tirado-Rives, J.; Jorgensen, W. L. *J. Med. Chem*. **2006**, 49, 5880–5884.
73. Nettles, J. H.; Jenkins, J. L.; Williams, C.; Clark, A. M.; Bender, A.; Deng, Z.; Davies, J. W.; Glick, M. *J. Mol. Graph. Model*. **2007**, 26(3), 622.
74. Cramer, R. D.; Jilek, R. J.; Andrews, K. M. *J. Mol. Graph. Model*. **2002**, 20, 447–462.
75. Jilek, R. J.; Cramer, R. D. *J. Chem. Inf. Comput. Sci*. **2004**, 44, 1221–1227.
76. Clark, T. *J. Mol. Graph. Model*. **2004**, 22, 519–525.
77. Ehresmann, B.; de Groot, M. J.; Alex, A.; Clark, T. *J. Chem. Inf. Comput. Sci*. **2004**, 44(2), 658–668.
78. Saunders, R. A.; Platt, J. A. *J. Phys. Org. Chem*. **2004**, 17(5), 431–438.
79. Clark, D. E. *J. Pharm. Sci*. **1999**, 88(8), 815–821.
80. Kelder, J.; Grootenhuis, P. D. J.; Bayada, D. M.; Delbressine, L. P. C.; Ploemen, J.-P. *Pharm. Res*. **1999**, 16(10), 1514.
81. Clark, D. E. *J. Pharm. Sci*. **1999**, 88, 807–814.
82. Palm, K.; Luthman, K.; Ungell, A. L.; Strandlund, G.; Artursson, P. *J. Pharm. Sci*. **1996**, 85(1), 32–39.
83. Palm, K.; Stenberg, P.; Luthman, K.; Artursson, P. *Pharm. Res*. **1997**, 14(5), 568.
84. Camenisch, G.; Folkers, G.; van de Waterbeemd, H. *Pharm. Acta Helv*. **1996**, 71, 309–327.
85. Veber, D. F.; Johnson, S. R.; Cheng, H. Y.; Smith, B. R.; Ward, K. W.; Kopple, K. D. *J. Med. Chem*. **2002**, 45(12), 2615–2623.
86. Palm, K.; Luthman, K.; Ungell, A.-L.; Strandlund, G.; Artursson, P. *J. Pharm. Sci*. **1996**, 85(1), 32–39.
87. Winiwarter, S.; Ax, F.; Lennernäs, H.; Hallberg, A.; Pettersson, C.; Karlén, A. *J. Mol. Graph. Model*. **2003**, 21(4), 273.
88. Stenberg, P.; Norinder, U.; Luthman, K.; Artursson, P. *J. Med. Chem*. **2001**, 44(12), 1927–1937.
89. Becker, O.; Mackerell, A. J.; Roux, B.; Watanabe, M. *Computational Biochemistry and Biophysics*. Marcel Dekker, New York, 2001.
90. Roux, B.; Simonson, T. *Biophys. Chem*. **1999**, 78(1–2), 1.
91. Eisenberg, D.; McClachlan, A. *Nature* **1986**, 319, 199–203.
92. am Busch, M.; Lopes, A.; Amara, N.; Bathelt, C.; Simonson, T. *BMC Bioinf*. **2008**, 9(1), 148.
93. Whitehead, C. E.; Breneman, C. M.; Sukumar, N.; Ryan, M. D. *J. Comput. Chem*. **2003**, 24, 512–529.

94. Sukumar, N.; Breneman, C. M. QTAIM in drug discovery and protein modeling. In *The Quantum Theory of Atoms in Molecules: From Solid State to DNA and Drug Design*, Matta, C. F.; Boyd, R. J., Eds. Wiley-VCH, Weinheim, Germany, 2007, pp. 471–498.
95. Bader, R. F. W. *Atoms in Molecules: A Quantum Theory*. Oxford University Press, Oxford, UK, 1990.
96. Matta, C. F.; Boyd, R. J. *The Quantum Theory of Atoms in Molecules: From Solid State to DNA and Drug Design*. Wiley-VCH, Weinheim, Germany, 2007.
97. Politzer, P.; Murray, J. S.; Peralta-Inga, Z. *Int. J. Quantum Chem*. **2001**, 85(6), 676–684.
98. Politzer, P.; Truhlar, D. G. *Chemical Applications of Atomic and Molecular Electrostatic Potential*. Plenum Press, New York, 1981.
99. Murray, J. S.; Politzer, P. *Theor. Comput. Chem*. **1998**, 5, 198–202.
100. Murray, J. S.; Politzer, P.; Famini, G. R. *Theochem J. Mol. Struct*. **1998**, 454(2–3), 299–306.
101. Fukui, K. *Science* **1987**, 218, 747–754.
102. Parr, R. G.; Yang, W. *J. Am. Chem. Soc*. **1984**, 106, 4049–4050.
103. Song, M.; Breneman, C. M.; Bi, J.; Sukumar, N.; Bennett, K. P.; Cramer, S.; Tugcu, N. *J. Chem. Inf. Comput. Sci*. **2002**, 42, 1347–1357.
104. Mazza, C. B.; Sukumar, N.; Breneman, C. M.; Cramer, S. *Anal. Chem*. **2001**, 73(22), 5457–5461.
105. Doytchinova, I. A.; Walshe, V.; Borrow, P.; Flower, D. R. *J. Comput.-Aided Mol. Des*. **2005**, 19, 203–212.
106. Grantham, R. *Science* **1974**, 185, 862–864.
107. Schneider, G.; Wrede, P. *Biophys J*. **1994**, 66, 335–344.
108. Kono, H.; Sarai, A. *Proteins* **1999**, 35, 114–131.
109. Lawrence, C. E.; Altschul, S. F.; Boguski, M. S.; Liu, J. S.; Neuwald, F.; Wootton, J. C. *Science* **1993**, 262(5131), 208–214.
110. Neuwald, A. F.; Liu, J. S.; Lawrence, C. E. *Protein Sci*. **1995**, 4, 1618–1632.
111. Stormo, G. D.; Hartzell, G. W., III. *Proc. Natl. Acad. Sci. USA* **1989**, 86, 1183–1187.
112. Bailey, T. L.; Elkan, C. Fitting a mixture model by expectation maximization to discover motifs in biopolymers. In *Proc. 2nd Int. Conf. Intell. Sys. Molec. Biol.*1994, pp. 28–36.
113. Cardon, L. R.; Stormo, G. D. *J. Mol. Biol*. **1992**, 223, 159–170.
114. Lawrence, C. E.; Reilly, A. A. *Proteins* **1990**, 7, 41–51.
115. Lawrence, C. E.; Reilly, A. A. *J. Am. Statist. Assoc*. **1996**, 91(433), 76–85.
116. Sukumar, N.; Lawrence, C. E.; Breneman, C. M.; Bennett, K. P.; Vitol, I. Fuzzy bar code representations of DNA–protein interactions. Presented at the 226th ACS National Meeting, New York, 2003.

117. Sukumar, N.; Lawrence, C. E.; Breneman, C. M.; Bennett, K. P.; Vitol, I. Dixel modeling of gene expression. Presented at the 227th ACS National Meeting, Anaheim, CA, 2004.

118. Duda, R.; Hart, P. *Pattern Classification and Scene Analysis*. Wiley, New York, 1973.

119. Rosipal, R.; Trejo, L. J. *Machine Learn. Res*. **2001**, 2, 97–123.

120. Masek, B. B.; Merchant, A.; Matthew, J. B. *J. Med. Chem*. **1993**, 36, 1230–1238.

121. Burden, F. R. *J. Chem. Inf. Comput. Sci*. **1989**, 29, 225–227.

122. Burden, F. R. *Quant. Struct. Act. Relat*. **1997**, 16, 309–314.

123. Pearlman, R. S.; Smith, K. M. *3D-QSAR in Drug Design*. Kluwer/Escom, Dordrecht, The Netherlands, 1998, Vol. 2, pp. 339–353.

124. Pearlman, R. S.; Smith, K. M. *J. Chem. Inf. Comput. Sci*. **1999**, 39, 28–35.

125. Stanton, D. T. *J. Chem. Inf. Comput. Sci*. **1999**, 39(1), 11–20.

126. Sundling, C. M.; Sukumar, N.; Zhang, H.; Embrechts, M. J.; Breneman, C. M. *Rev. Comput. Chem*. **2006**, 22, 295–329.

127. Cruciani, G.; Watson, K. *J. Med. Chem*. **1994**, 37, 2589–2601.

128. Pastor, M.; Cruciani, G.; McLay, I.; Pickett, S.; Clementi, S. *J. Med. Chem*. **2000**, 43, 3233–3243.

129. Crivori, P.; Cruciani, G.; Carrupt, P.-A.; Testa, B. *J. Med. Chem*. **2000**, 43, 2204–2216.

130. Amat, L.; Carbo-Dorca, R.; Ponec, R. *J. Comput. Chem*. **1998**, 19, 1575–1583.

131. Carbo, R. *Molecular Similarity and Reactivity: From Quantum Chemical to Phenomenological Approaches*. Kluwer Academic, Amsterdam, 1995.

132. Carbo, R.; Arnau, J.; Leyda, L. *Int. J. Quantum Chem*. **1980**, 17, 1185–1189.

133. Carbo, R.; Besalu, E.; Amat, L.; Fradera, X. *J. Math. Chem*. **1995**, 18, 237–246.

134. Carbo, R.; Calabuig, B. *J. Mol. Struct*. (*Theochem*) **1992**, 254, 517–531.

135. Carbo, R.; Calabuig, B.; Vera, L.; Besalu, E. *Adv. Quantum Chem*. **1994**, 25, 253–313.

136. Carbo-Dorca, R.; Mezey, P. G. *Advances in Molecular Similarity*. JAI Press, Greenwich, CT, 1996.

137. Carbo-Dorca, R.; Robert, D.; Amat, L.; Girones, X.; Besalu, E. *Molecular Quantum Similarity in QSAR and Drug Design*. Springer-Verlag, Berlin, 2000.

138. O'Brien, S. E.; Popelier, P. L. A. *Can. J. Chem*. **1999**, 77, 28–36.

139. O'Brien, S. E.; Popelier, P. L. A. *J. Chem. Inf. Comput. Sci*. **2001**, 41, 764–775.

140. O'Brien, S. E.; Popelier, P. L. A. *Perkin Trans*. **2002**, II, 478–483.

141. Ponec, R.; Amat, L.; Carbo-Dorca, R. *J. Comput.-Aided Mol. Des*. **1999**, 13, 259–279.

142. Ponec, R.; Amat, L.; Carbo-Dorca, R. *J. Phys. Org. Chem*. **1999**, 12, 447–454.

143. Popelier, P. L. A. *J. Phys. Chem*. **1999**, 103, 2883–2890.

144. Popelier, P. L. A.; Chaudry, U. A.; Smith, P. J. *Perkin Trans*. **2002**, II, 1231–1237.

145. Leherte, L.; Meurice, N.; Vercauteren, D. P. *J. Chem. Inf. Comput. Sci*. **2000**, 40, 816–832.

146. Todeschini, R.; Gramatica, P. *Quant. Struct.-Act. Relat*. **1997**, 16, 120–125.

147. Todeschini, R.; Gramatica, P.; Provenzani, R. *Chemometr. Intell. Lab. Syst*. **1995**, 27, 221–229.

148. Todeschini, R.; Lasagni, M.; Marengo, E. *J. Chemometr*. **1994**, 8, 263–272.

149. Consonni, V.; Todeschini, R.; Pavan, M. *J. Chem. Inf. Comput. Sci*. **2002**, 42, 682–692.

150. Consonni, V.; Todeschini, R.; Pavan, M.; Gramatica, P. *J. Chem. Inf. Model.*, **2002**, 42, 693–705.

151. Gasteiger, J.; Sadowski, J.; Schuur, J.; Selzer, P.; Steinhauer, L.; Steinhauer, V. *J. Chem. Inf. Comput. Sci*. **1996**, 36, 1030–1037.

152. Holm, L.; Sander, C. *Science* **1996**, 273, 595–603.

153. Holm, L.; Sander, C. DALI. http://ekhidna.biocenter.helsinki.fi/dali_server.

154. Daras, P.; Zarpalas, D.; Axenopoulos, A.; Tzovaras, D.; Strintzis, M. G. *IEEE/ACM Trans. Comput. Biol. Bioinf.* **2006**, 3(3).

155. Morris, R. J.; Kahraman, A.; Funkhouser, T.; Najmanovich, R.; Stockwell, G.; Glaser, F.; Laskowski, R.; Thornton, J. M. In *Binding Pocket Shape Analysis for Protein Function Prediction*, LASR Workshop on Quantitative Biology, Shape Analysis, and Wavelets, Leeds, UK, 2005.

156. Morris, R. J.; Najmanovich, R. J.; Kahraman, A.; Thornton, J. M. *Bioinformatics* **2005**, 21(10), 2347–2355.

157. Ritchie, D. W.; Kemp, G. J. L. *J. Comput. Chem*. **1999**, 4, 383–395.

158. Morris, R. J. *J. Mol. Graph. Model.* **2006**, 24(5), 356–361.

159. Nebel, J. C.; Herzyk, P.; Gilbert, D. R. *BMC Bioinf*. **2007**, 8, 321.

160. Sims, G. E.; Kim, S.-H. *Proc. Natl. Acad. Sci. USA* **2005**, 102(12), 618–621.

161. Mezei, M. *Protein Eng*. **2003**, 16, 713–715.

162. Rogen, P.; Bohr, H. *Math. Biosci*. **2003**, 182, 167–181.

163. Luo, Q. Design, *Development and Utilization of Novel Molecular Descriptors and Machine Learning Methods*. Rensselaer Polytechnic Institute, Troy, NY, 2006.

164. Cramer, S. M. In *Prediction of Protein Affinity in HIC Systems Using State-of-the-Art Modeling Techniques*. New York Academy of Sciences, Brooklyn, NY, 2005.

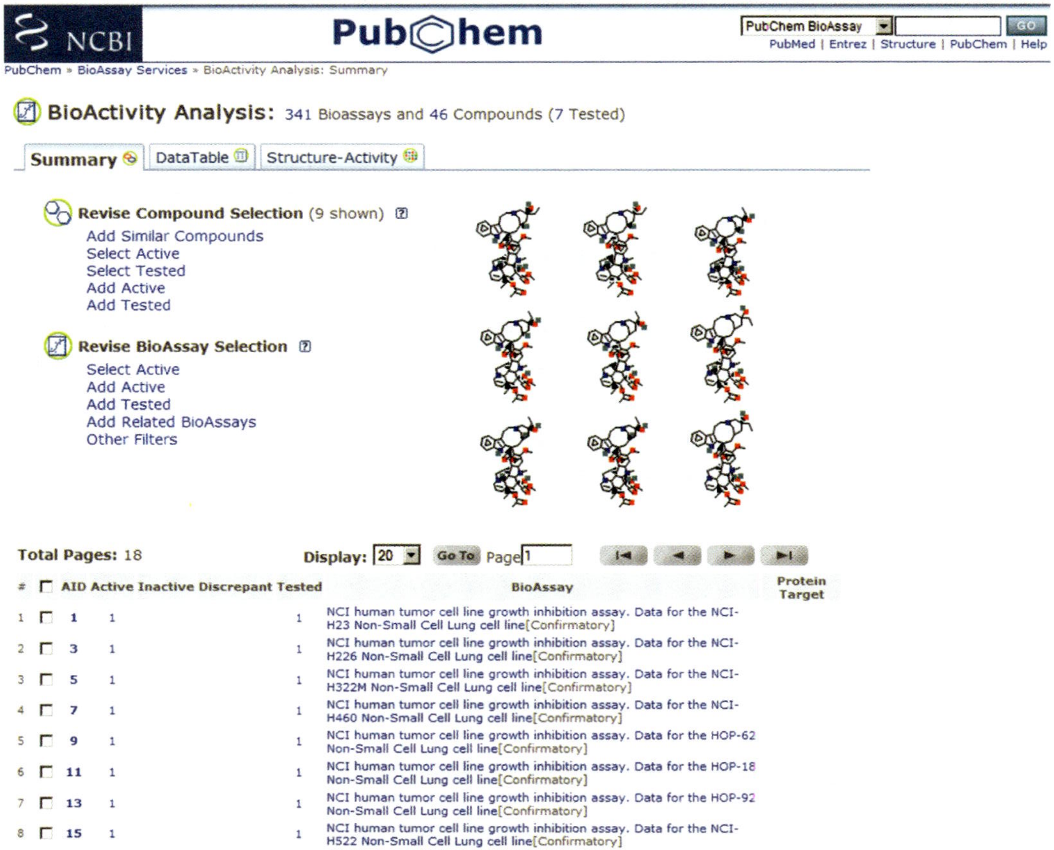

Chapter 1, Figure 3 Partial view of a PubChem bioactivity analysis display for the 46 compounds that have the same parent compound connectivity to CID 13342.

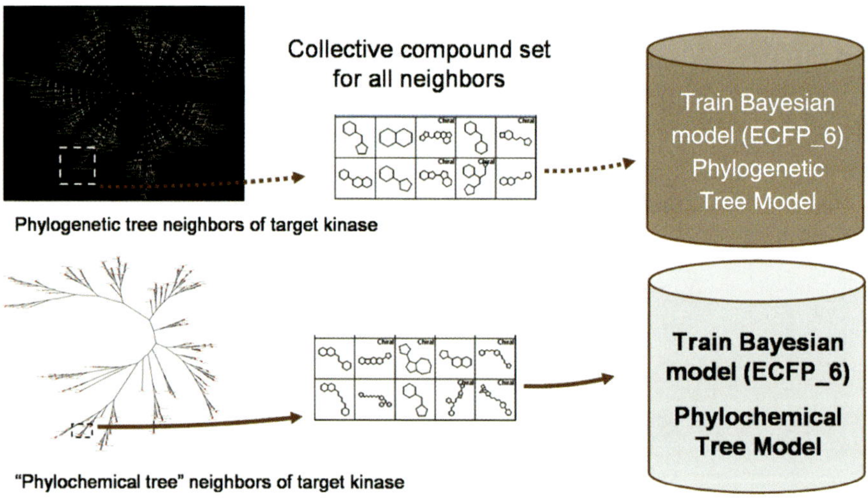

Chapter 2, Figure 5 Workflow for chemogenomics-driven lead finding. Based on phylogeny of a new protein target to other known targets, inhibitor compounds can be found. In this example, the position of a kinase on a sequence-based phylogenetic tree is used to find its neighboring branch nodes. The compound inhibitors of the tree neighbors are aggregated and used to train a Bayesian classification model, which is then used to rank a chemical library for potential leads. Targets with at least 1 or 2 tool compounds may be placed on a phylochemical tree by profiling the tool compounds in kinase assay panels or by small-molecule affinity chromatography to determine common "off-targets". These off-targets enable placement on the kinase-of-interest on a phylochemical tree. The workflow then follows that of the previous approach.

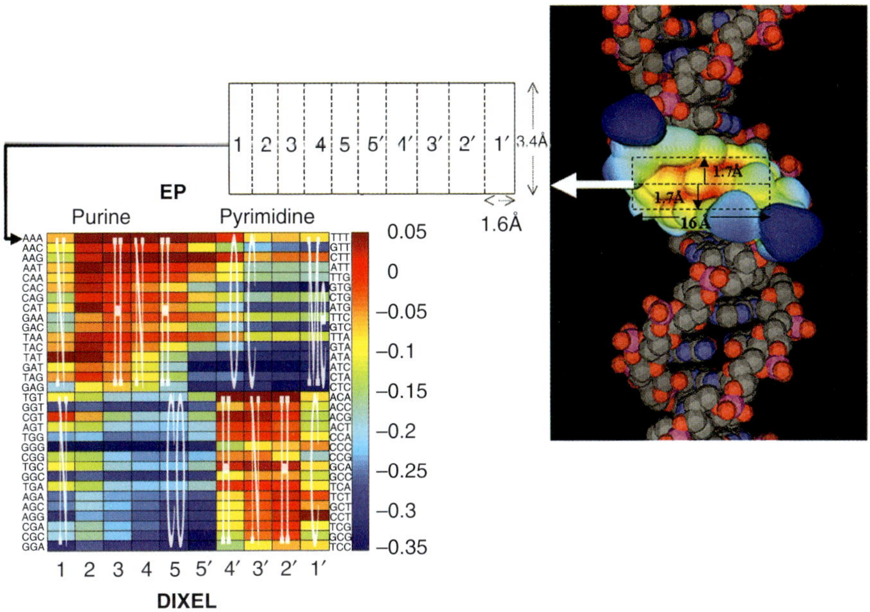

Chapter 5, Figure 1 Mapping from the DNA major groove surface to a Dixel representation. Each base pair is represented by a rectangular grid of surface pixels. The resulting library of base pair triples is employed to reconstruct the DNA sequence based on the surface electron density properties of the central base pair of each triplet.

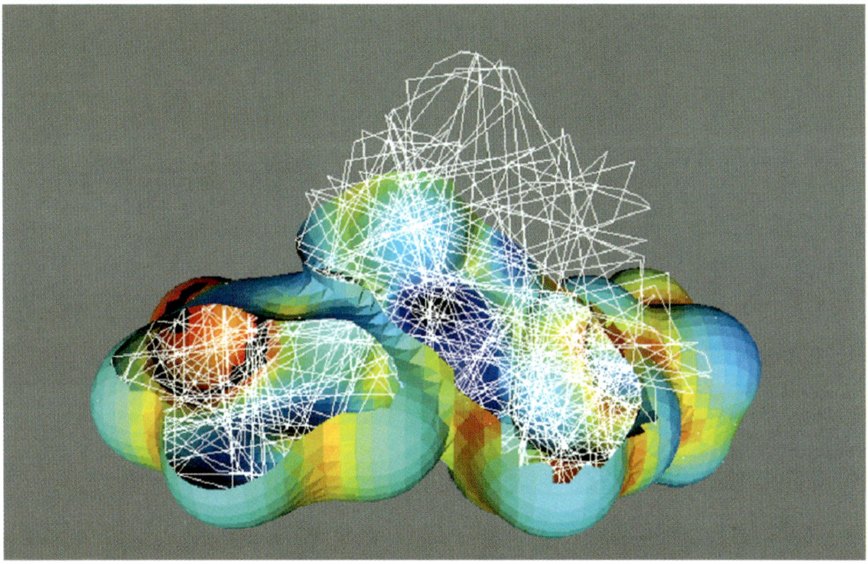

Chapter 5, Figure 4 Property-encoded ray tracing from a sparse 750-segment PEST run, with part of the molecular van der Waals surface cut away to show the ray bounces.

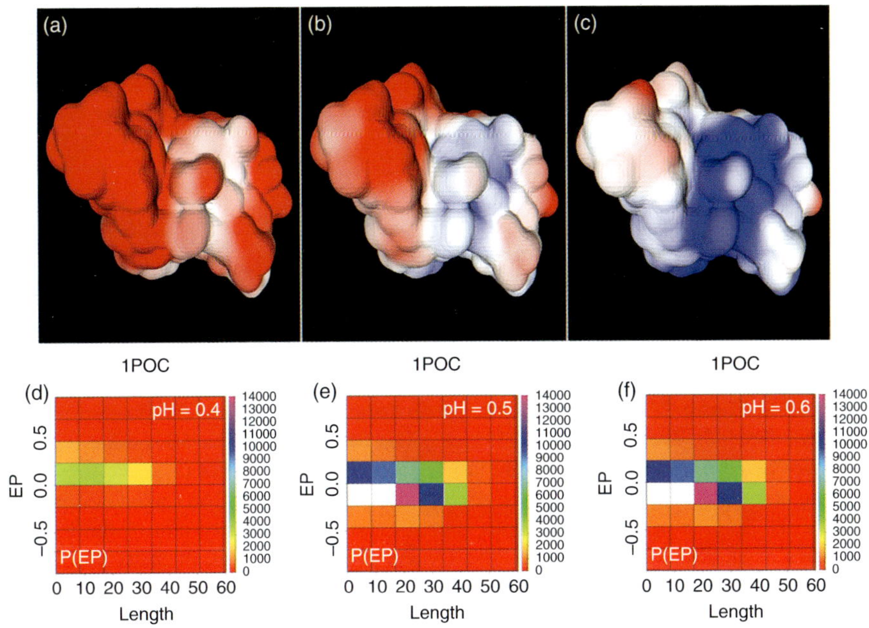

Chapter 5, Figure 6 EP surface profiles from PPEST for the protein 1POC at three different values of pH [(a) pH 4.0, (b) pH 6.0, and (c) pH 8.0], and corresponding shape/property histograms [at (d) pH 4.0, (e) pH 6.0, and (f) pH 8.0].

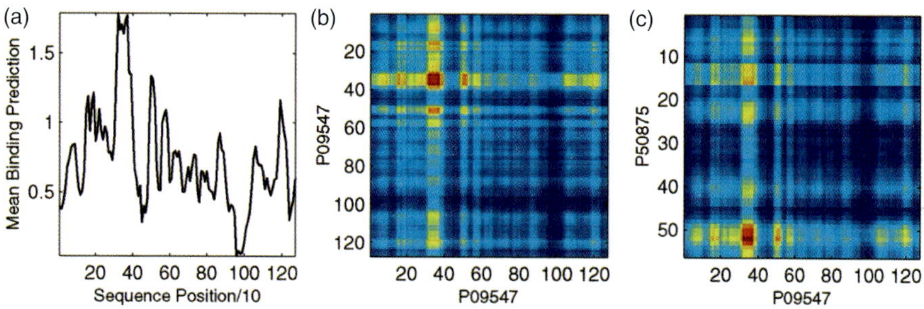

Chapter 6, Figure 2 Protein-domain predictions. Plots showing the domain predictions for P09547. The x-axis of each plot gives the position of a 50-residue window moved 10 residues at a time across the full sequence of P09547. Plot (a) shows the mean binding activity of the windows along with x-axis with the other 1681 windows considered in the domain prediction example; plot (b) shows an intensity plot of the binding activities of all pairs of windows in P09547; and plot (c) shows an intensity plot of the binding activities of all pairs of windows in P09547 and P50875 (a known binder). In (b and c), red denotes activity and blue denotes inactivity.

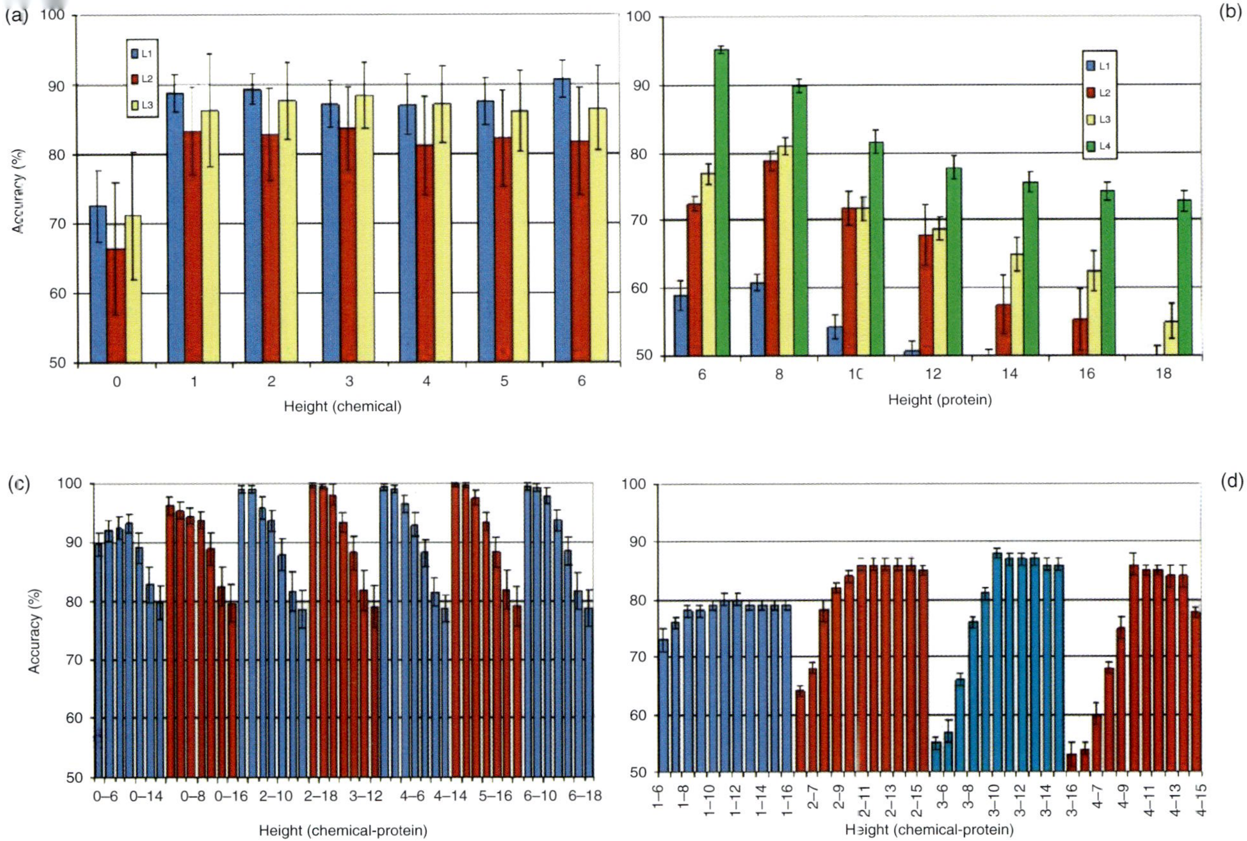

Chapter 6, Figure 3 Fivefold cross-validation prediction accuracies for EC numbers. Signature kernels for various chemical, proteins, and signature heights are used. (a) Accuracies of the signature kernel for the first three EC levels (L1 to L3) using only chemical reactions. (b) Accuracies of the signature kernel for the four EC levels (L1 to L4) using only protein sequences. (c) Accuracies of the signature product kernel for the EC class 1.1.1.1. (d) Accuracies for the KEGG drug–target data set using the signature product kernel.

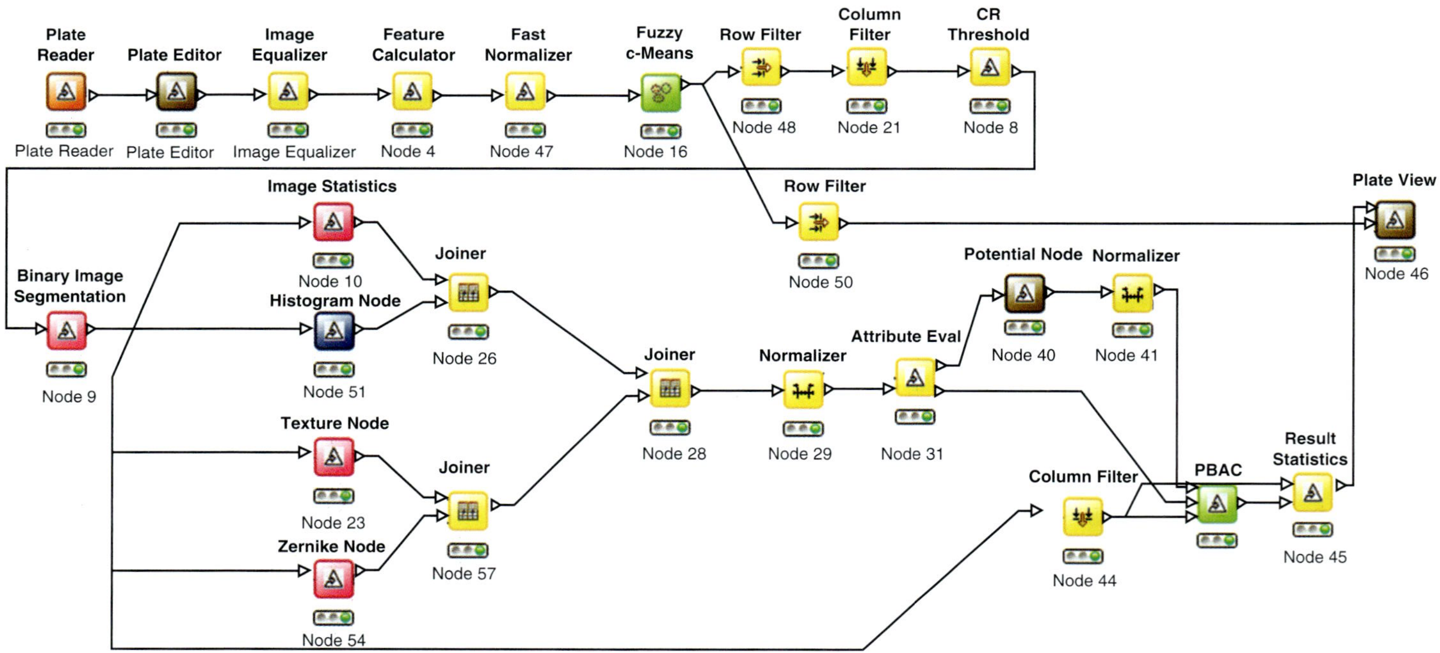

Chapter 7, Figure 11 Workflow demonstrating the analysis of cell assay images by various image-processing nodes.

165. Sukumar, N.; Breneman, C. M.; Sundling, C. M. New tools for virtual high-throughput screening. Presented at the ACS National Meeting, San Francisco, 2006.

166. Morrison, C.; McCallum, S.; Godawat, R.; Moore, J. A.; Garde, S.; Cramer, S. M. Investigation of chemically selective displacers using robotic high throughput screening, SPR, NMR and MD simulations. Presented at the ACS National Meeting, Boston, 2007.

167. Garde, S.; Hummer, G.; García, A. E.; Pratt, L. R.; Paulaitis, M. E. *Phys. Rev. E* **1996**, 53(5), R4310–R4313. Garde, S.; Hummer, G.; Paulaitis, M. E.; García, A. E. "Hydration of biological macromolecules: From small solutes to proteins and nucleic acids", in *Proc. 1996 MRS Fall Meet. Symp. Stat. Mech. Phys. Biol.*, eds. D. Wirtz, T. C. Halsey, **1997**, 463, 21.

168. Das, S.; Breneman, C. M.; Sukumar, N.; Ryan, M. D. TAE augmented scoring functions: application to enzymatic and nonenzymatic proteins. Presented at the 235th ACS National Meeting, New Orleans, LA, 2008.

169. Oloff, S.; Zhang, S.; Sukumar, N.; Breneman, C.; Tropsha, A. *J. Chem. Inf. Model*. **2006**, 46(2), 844–851.

170. Golbraikh, A.; Tropsha, A. *J. Mol. Graph. Model*. **2002**, 20(4), 269–276.

171. Bender, A.; Mussa, H. Y.; Glen, R. C. *J. Chem. Inf. Comput. Sci*. **2004**, 44, 170–178.

172. Sundling, C. M.; Breneman, C. M.; Embrechts, M. J.; Huang, C.; Wu, X.; Sukumar, N. Testing the limits of a QSAR model: How many cases are actually needed to develop a reliable predictive model? Presented at the 235th ACS National Meeting, New Orleans, LA, 2008.

173. Ryan, D.; McLellan, M.; Breneman, C. M. QSAR model stability: How much information is in the data? Presented at the 235th ACS National Meeting, New Orleans, LA, 2008.

174. Audry, E.; Dubost, J.-P.; Colleter, J.-C.; Dallet, P. *Eur. J. Med. Chem. Chim. Ther*. **1986**, 21, 71.

175. Heiden, W.; Moeckel, G.; Brickmann, J. *J. Comput.-Aided Mol. Des*. **1993**, 7, 503–514.

176. Parr, R. G.; Yang, W. *Density-Functional Theory of Atoms and Molecules*. Oxford University Press, New York, 1989.

CHAPTER 6

GRAPHS: FLEXIBLE REPRESENTATIONS OF MOLECULAR STRUCTURES AND BIOLOGICAL NETWORKS

MILIND MISRA
Advanced Device Technologies, Sandia National Laboratories, Albuquerque, New Mexico

SHAWN MARTIN
Computer Science and Informatics, Sandia National Laboratories, Albuquerque, New Mexico

JEAN-LOUP FAULON
Advanced Device Technologies, Sandia National Laboratories, Albuquerque, New Mexico

1 INTRODUCTION

The past two decades have seen a large accumulation of biological sequences and chemical compounds in many publicly available databases. For a long time, the two communities of bioinformatics and cheminformatics have developed in parallel, working largely with sequence data alone or mainly in the chemical space, respectively. As the need to study biological networks has increased, however, a concurrent need to develop tools and algorithms capable of handling the combined sequence and chemical space has arisen. We present here a graph-based technique, *molecular signature*, which is sufficiently adaptable to permit a combined description, for high-throughput analyses, of sequences and chemicals.

Computational Approaches in Cheminformatics and Bioinformatics, First Edition.
Edited by Rajarshi Guha and Andreas Bender.

We first demonstrate the ability of the technique to describe chemicals and sequences and predict chemical properties and activities and proteins functions and interactions. We next combine chemical and sequence descriptions to represent protein–chemical complexes, and we demonstrate how the combined representation can be used to infer metabolic networks and predict drug–target interactions.

In Section 2 we introduce the signature descriptor and describe briefly its applications in chemistry. After a formal description of signature, we demonstrate that molecular signatures encompass information contained in many other molecular descriptors and the application of signatures in quantitative structure activity–property relationship analyses. In Section 3 we present applications of signatures in biology. It is shown that signatures can be used to construct kernel functions for use in support vector machine learning to predict protein–protein interactions. Finally, in Section 4, molecular and protein signatures are merged with the goal of predicting enzyme–metabolite interactions and drug–target binding.

Classical cheminformatics techniques can predict drugs binding to a given target, and bioinformatics tools can predict enzymes catalyzing a given metabolite. However, in either case, training sets with a sufficient number of drugs for a given target, or a sufficient number of enzymes for a given metabolic reaction, are required. It is demonstrated that by using signature product between chemicals and sequences, machine learning techniques can be trained on heterogeneous data sets containing chemicals *and* proteins. Results indicate that it is possible to predict enzymes catalyzing reactions, or drugs binding to targets, even for enzymes and reactions, or targets and drugs, not present in the training sets.

2 CHEMICAL SIGNATURE: MOLECULAR DESIGN AND QSAR/QSPR

The properties of a molecule can be characterized by representing its two-dimensional (2D) structure as a graph with vertices (atoms) and edges (bonds) and by using operators (descriptors) on this graph. These operators may be nonempirical, often having only qualitative meaning at the outset of their introduction, with more refined interpretation following during subsequent analysis.[1] The numerical values resulting from the operation of a given descriptor on a graph may be used as independent variables in quantitative structure–activity or structure–property relationship analyses (QSAR/QSPR).

There are numerous descriptors available for QSAR/QSPR studies, and the availability of commercial packages with QSAR/QSPR capabilities has made selection of the proper descriptor set research dependent.[2] Although the three-dimensional (3D) conformation of a molecule is more useful for spatial relationships of atoms and bonds than the 2D chemical graph, the topological (2D) descriptors have been shown in some studies to perform as well as 3D descriptors.[3] Since the graph of a molecule itself contains limited information relative to the available independent variables, many of the descriptors in a set are highly correlated. In addition, there is usually no one-to-one mapping between a descriptor and a molecular property. In past work, Randić and Basak ask: "Do we need additional descriptors although hundreds of molecular descriptors ... are already available? ... How can we establish whether the available molecular descriptors suffice for a complete characterization of molecules for QSPR and QSAR?"[4] A step toward addressing this important issue is the generation of a finite set of descriptors that are not highly correlated and form a complete set from which all other descriptors can be calculated. In this section we introduce the concept of a molecular signature, demonstrate how topological indices can be calculated from a signature, and apply signatures in QSAR/QSPR analyses.

The Signature Descriptor

The *signature* is a systematic codification system over an alphabet of atom types. It describes the extended valence or neighborhood of the atoms of a molecule. This concept was first presented and applied in the context of structure elucidation,[5] and later defined for acyclic compounds and used in QSAR analysis.[6] This introduction of the signature descriptor necessitates the definition of some terminology and notation.

Molecular Graph A molecule is represented by a graph $G = (V_G, E_G, C, c_G(\cdot))$, where the elements of the matrix of vertices, V_G, are atoms and the elements of the matrix of edges, E_G, are bonds. The atoms of a molecular graph are colored by a set of atom types, C, which, for instance, can be the set of elements of the periodic table or any set of atom types provided by a molecular force field. The function $c_G(\cdot)$ associates an atom in G to an atom type in C. Every atom type has a valence, which is the number of covalent bonds that can be formed with this atom. A *molecular graph* is a graph representing a molecule which is not necessarily saturated. Formally, a molecular graph $G = (V_G, E_G, C, c_G(\cdot))$ is an undirected graph colored with the function $c_G(\cdot)$ over the elements of C satisfying the inequality:

$$\forall x \in V_G^{\circ}(x) \leq \text{valence}(c_G(x)) \tag{1}$$

Signature of an Atom Let G be a molecular graph and let x be an atom of G. The signature for height h about x, ${}^h\sigma_G(x)$, is a canonical representation of the subgraph of G containing all atoms that are at distance h from x. This canonical representation takes the form of a tree and is constructed using the following five-step procedure (Figure 1):

Step 1. A subgraph ${}^hG(x)$ is extracted from G containing all atoms that are at distance h from x and all bonds between these atoms.

Step 2. The vertices of ${}^hG(x)$ are labeled in a canonical order with atom x labeled 1.

Step 3. A tree spanning all the edges of ${}^hG(x)$ is constructed. The root of the tree is atom x itself. The first layer of the tree is composed of the neighbors of x, and the second layer is composed of the neighbors of the vertices of the first layer except atom x. To each vertex added to the tree one associates the color and the canonical label of the corresponding atom. The construction proceeds one layer at a time up to layer h. Assume that the tree has been constructed up to layer $k < h$; layer $k+1$ is constructed considering each vertex y of layer k. Let z be a neighbor of y in G. Vertex z and edge $[y, z]$ are added to layer $k+1$ if the edges $[y, z]$ or $[z, y]$ are not already present in the tree. The neighbors of y present in layer $k+1$ are sorted in decreasing lexicographic order using the colors and the canonical labels of the corresponding atoms. As shown in Figure 1, in the first layer, C,3, appears before C,2. Note that while a given vertex z (such as C,7) may appear several times in the tree since it can be the neighbor of several vertices present in the preceding layer, a given edge may appear only once.

Step 4. Once the tree has been constructed up to layer h, all canonical labels that appear only once are removed. The remaining labels are renumbered in the order in which they appear while reading the tree in a depth-first order.

Step 5. The signature is written by reading the tree in a depth-first order and printing the character '(' each time an edge from parent to child is read, the character ')' when the edge is read from child to parent, and the vertex color followed by a label if the vertex appears more than once in the tree. In Figure 1, there is only one atom type—sp3 hybridized carbon—whose "color" in the signature is indicated by the letter C.

Signature of a Molecule By the definition above, the signature of an atom can be viewed as a string of characters over the alphabet C of atom types. Note that for a given height h, the list of all possible atomic signatures (${}^h\Sigma$), although large, is of finite size. Consequently, any

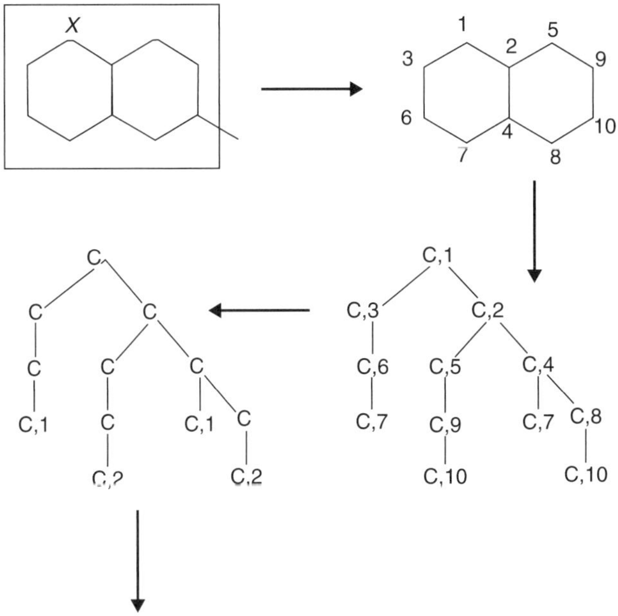

${}^{0}\sigma(X) = \mathrm{C}$

${}^{1}\sigma(X) = \mathrm{C\ (C\ C)}$

${}^{2}\sigma(X) = \mathrm{C\ (C\ (C)\ C(C\ C))}$

${}^{3}\sigma(X) = \mathrm{C\ (C\ (C\ (C,1))\ C(C\ (C)\ C\ (C,1\ C)))}$

${}^{4}\sigma(X) = \mathrm{C\ (C\ (C\ (C,1))\ C(C\ (C\ (C,2))\ C\ (C,1\ C(C,2))))}$

FIGURE 1 Atomic signature: the five-step procedure for computing the atomic signature of atom x in 2-methyldecahydronaphthalene. (1) The subgraph containing all atoms at distance 4 from atom x is extracted. (2) The subgraph is canonized atom x having label 1. (3) A tree spanning all edges of the subgraph is constructed. (4) All labels appearing only once are removed and the remaining labels are renumbered in the order they appear. (5) The signature is printed reading the tree in a depth-first order. The figure shows atomic signatures from $h = 0$ to $h = 4$.

molecule of the chemical universe can be represented by its coordinates in a vector space where the base vectors are the distinct atomic signatures. The signature of a molecule is thus defined as the linear combination of its atomic signatures:

$$ {}^{h}\sigma(G) = \sum_{x \in V_G} {}^{h}\sigma_G(x) = \alpha(G, {}^{h}\Sigma) \tag{2} $$

where ${}^h\Sigma$ is the alphabet of all possible atomic signatures of height h, and $\alpha(G, {}^h\Sigma)$ is the vector of occurrence numbers of atomic h-signatures of graph G.

Signature of a Bond Let $G = (V_G, E_G, C, c_G(\cdot))$ be a molecular graph and let b be a bond/edge of E_G. Let $G - b = (V_G, E_G - b, C, c_G(\cdot))$ be the molecular graph in which the bond b has been removed. The h-signature of b is defined as follows:

$$ {}^h\sigma(b) = {}^h\sigma(G) - {}^h\sigma(G - b) \tag{3} $$

Signature of a Reaction Let $B = (V_B, E_B, C, c_G(\cdot))$ and $E = (V_E, E_E, C, c_G(\cdot))$ be two molecular graphs representing the reactants and products of the reaction *R:* $B \to E$. Note that signatures can be computed on graphs that are not necessarily connected; hence B and E can both be composed of several molecules. The h-signature of reaction R is given by the equation

$$ {}^h\sigma(R) = {}^h\sigma(E) - {}^h\sigma(B) \tag{4} $$

Computational Space and Time Complexity of Signature In the following, G is a molecular graph of n atoms, x is an arbitrary atom of G, and h is the height of the signatures considered. Let D be the diameter of G, which is the largest distance between all pairs of atoms in G. Note that a priori D may be equal to n. ${}^hG(x)$ is the subgraph composed of all vertices of G that are at most h-distant from x, ${}^h\sigma_G(x)$ is the atomic h-signature of x, and ${}^h\sigma(G)$ is the molecular h-signature of G. The following five statements are proven in Faulon et al.[7]:

1. The number of characters used to print ${}^h\sigma_G(x)$ scales $O(n)$.
2. The numbers of characters used to print ${}^h\sigma(G), {}^h\sigma(b)$, and ${}^h\sigma(R : B \to E)$ scale $O(n^2)$.
3. Once initialized, the adjacency matrix of G can be constructed in $O(n)$ steps from ${}^{D+1}\sigma(G)$.
4. ${}^{D+1}\sigma(G)$ is a canonical representation of G.
5. ${}^h\sigma(G)$ can be computed with a number of steps bounded by a polynomial function in n. This is a theoretical result relying on transforming molecular graphs into simple graphs as described elsewhere.[8] A practical algorithm to canonize signature is given by Faulon et al.[9] and can also be downloaded at http://jfaulon.wikispaces.com/.

Computing Topological Indices from Signature

In the following, G is the graph of a hydrogen-suppressed covalent molecule with n atoms and diameter D. The molecule has known signatures up to height h, ${}^h\sigma(G) = \alpha(G, {}^h\Sigma)$, where ${}^h\Sigma$ is the alphabet of all atomic signatures of height h and $\alpha(G, {}^h\Sigma)$ is the vector of occurrence numbers of atomic h-signatures of molecule G. As stated above, the adjacency matrix of the entire molecular graph can be computed from the height $h = D+1$ molecular signature. Consequently, any topological index (TI) can be computed from a molecular signature of height $D+1$, simply because any TI can be computed from an adjacency matrix. A less trivial result is the minimum height for which a given TI can be computed. Table 1 summarizes the results for a subset of popular TIs used in QSAR and QSPR. As demonstrated by Faulon et al.,[7] for any chosen TI, there exists a height $h < D+1$ for which the index can be expressed as a dot product between two vectors, $\alpha(G, {}^h\Sigma)$, the vector of occurrence numbers of atomic h-signatures of molecule G, and $\mathrm{TI}(\mathrm{root}({}^h\Sigma))$, the vector of TI values computed for each root of the atomic h-signatures of ${}^h\Sigma$:

$$\mathrm{TI}(G) = \mathrm{constant} \cdot \alpha(G, {}^h\Sigma) \cdot \mathrm{TI}(\mathrm{root}({}^h\Sigma)) \tag{5}$$

Equation (5), illustrated in Table 1, demonstrates that 2D descriptors can be computed from signature. From this it follows that any existing QSAR

TABLE 1 List of the TIs Computable from the Molecular h-Signature

h	TIs
0	n: number of atoms; MF: molecular formula, MW: molecular weight; fragments of one atom.
1	MF and MW for hydrogen-suppressed graphs; m: number of bonds; μ: cyclomatic number; I: intrinsic state; mwc_1: molecular walk count of length 1; ${}^1\kappa, {}^1\kappa_\alpha$: shape indices of length 1; ${}^0\chi$, ${}^0\chi_v$: connectivity indices of length 0; p_1/w_1: Randić 145 shape index of length 1; fragments of diameter 1.
2	mwc_2; ${}^2\kappa, {}^2\kappa_\alpha$; F: Platt number; ${}^1\chi$, ${}^1\chi_v$; p_1/w_1; s_1: shell index of length 1; fragments of diameter 2.
L	mwc_l; ${}^l\kappa, {}^l\kappa_\alpha$; ${}^{l-1}\chi$, ${}^{l-1}\chi_v$; s_{l-1}; p_l/w_l; fragments of diameter l.
$l = D^a$	W, M, WW, H: Wiener-type indices; Balaban J index; I_E^P: information index.
$l = D+1$	S: E-state index; τ: total topological index for hydrogen-suppressed graphs; twc: total walk count; ω, $\omega\omega$: detour-type indices; Wt: all-path-based Wiener index.

[a] D represents the diameter of the moleculer graph.

or QSPR equation involving 2D descriptors can be replaced by an equivalent equation involving occurrence numbers of atomic signatures (e.g., substructural fragments). An example of such substitution for boiling-point prediction is given by Faulon et al.[7]

Signature in QSAR/QSPR

We now demonstrate the utility and robustness of the signature descriptor in two QSAR/QSPR studies. The first data set comes from a biological source and measures the activity for 121 compounds used as HIV-1 protease inhibitors. The much larger second data set is for the octanol/water partition coefficient (log P) of 12,988 compounds. The results of the models developed using signature are compared with the results of models developed using descriptors from Molconn-Z (version 3.50), a commercially available program from eduSoft, LC.[10]

Molconn-Z descriptors include (among others): connectivity indices, shape indices, information indices, fragment counts, and electrotopological states.[10,11] When using atomic signatures as descriptors, the occurrence number of a particular atomic signature is used as the descriptor value. (The occurrence number is the number of times a particular atomic signature appears in a given molecule's signature set ${}^{h}\Sigma$.) The overall number of unique atomic signatures in the data set is not known a priori and is a function of both the height used and the size of the training set. The QSAR/QSPR studies described here use atomic signatures of height 0 (0-signature), height 1 (1-signature), and height 2 (2-signature).

For both the Molconn-Z models and the signature-based models, the procedure to determine the overall number of descriptors available involved determining whether two or more descriptor vectors (the elements of the vectors are the descriptor values for each compound in the training set) were equivalent (i.e., had a pairwise correlation coefficient of unity). When this occurred, the descriptor that had the lowest overall absolute pairwise correlation coefficient with the rest of the descriptors in the training set was kept and the other descriptor was discarded. All models were of the linear form, and parameterization of the models was performed using forward-stepwise regression.[12] Briefly, forward-stepwise regression is a technique used for multiple linear regression wherein independent variables (here descriptors) are added to the model in a systematic way. In this work, variables were inserted into the model by choosing the descriptor with the largest partial correlation coefficient of the variables not currently in the model.

TABLE 2 QSAR Models for the HIV-1 Protease Inhibitors Data Set

Model	Number of Descriptors	Training Set (n = 121)		Test Set (n = 9) s
		r^2	s	
Molconn-Z	152	~1.0	3.038×10^{-6}	145.1
Molconn-Z[a]	31	0.951	0.410	0.575
0-signature	12	0.166	1.762	1.435
0-signature[a]	8	0.161	1.767	1.414
1-signature	121	0.991	0.186	701.9
1-signature[a]	18	0.915	0.503	0.633
2-signature	290	~1.0	3.819×10^{-7}	355.8
2-signature[a]	37	0.988	0.214	0.766

[a]Reduced models obtained by forward-stepwise regression.

QSAR for HIV-1 Protease Inhibitors The training set was created by selecting 121 HIV-1 protease inhibitors from six literature sources. The experimental activity reported was in units of $\text{pIC}_{50}(\text{pIC}_{50} = 9 - \log_{10}\text{IC}_{50})$ and spanned seven orders of magnitude. The complete list of compounds in this training set is given in the article by Faulon et al.[7] Nine compounds, with at least one compound from each of the six literature sources, were chosen for the test set. These compounds were not used in creating the models and were used only to evaluate the predictive ability of the models. The results are reported in Table 2. The models generated from all available Molconn-Z descriptors and from all available atomic height-2 signatures explain the training set almost perfectly ($r^2 = 1.0$). This is not surprising given that the number of parameters in both cases is greater than the number of observations. Models with such large parameter-to-observation ratios result in overfitting and, as indicated by the test set s values (root-mean-square error), are neither robust nor predictive. For each model, forward-stepwise regression was used to determine the number of parameters needed to generate the smallest test set s value. The resulting reduced models based on height-1 and height-2 atomic signatures compare well with those based on Molconn-Z descriptors, although the latter does predict the test set activities slightly better.

QSPR for Log P A large database of 12,988 compounds with their octanol/water partition coefficients (log P) was acquired from the Syracuse Research Corporation, and 123 compounds were selected for the test set. The results are listed in Table 3. Statistics for models using all available descriptors as well as for reduced models obtained by forward-stepwise regression are shown. (Note that the smallest test set

TABLE 3 QSPR Models for the Log *P* Data Set

Model	Number of Descriptors	Training Set (n = 12,865) r^2	Training Set (n = 12,865) s	Test Set (n = 123) s
Molconn-Z	192	0.881	0.629	0.722
Molconn-Z[a]	123	0.878	0.638	0.709
0-signature	16	0.635	1.103	1.124
1-signature	1161	0.922	0.508	0.770
1-signature[a]	265	0.914	0.536	0.632

[a]Reduced models obtained by forward-stepwise regression.

s value for the atomic height-0 signature model occurs when all 16 original signatures are used.) For this data set, the models developed from atomic height-1 signatures compare well with those developed from the Molconn-Z descriptors, with the signature-based reduced model giving better test set log P predictions.

Put to the test in two QSAR and QSPR studies, signature performed quite well compared with the Molconn-Z commercially available 2D descriptors. Comparable conclusions have also been drawn while evaluating HQSAR fragment-based holograms[13] with the CODESA descriptors to predict the binding of chemicals to estrogen receptors. Analogous to a molecular signature, a molecular hologram is composed of the substructural fragments constituting a molecule. Unlike signatures, holograms are stored in a bin array of predefined length, and thus the structures of the fragments are lost since several structurally different fragments may appear in the same bin. Yet it is surprising that both signatures and holograms, based solely on fragments, perform as well as hundreds of molecular descriptors encoding features as varied as molecular topology, geometry, electrostatics, and information content. A simple explanation of this result is that 2D descriptors can, in fact, be computed from molecular fragments (Table 1). As a consequence of this finding, and as demonstrated formally in Faulon et al.,[7] any existing QSAR or QSPR equation can be replaced with an equivalent equation involving occurrence numbers of substructural fragments (e.g., atomic signatures).

3 PROTEIN SIGNATURE: PREDICTION OF PROTEIN–PROTEIN INTERACTIONS

Proteome-wide prediction of protein–protein interactions is a fundamental biological problem. In this section we describe a general high-throughput method for predicting protein–protein interactions.[14] The method combines a signature-based description of proteins with experimental information that can be gathered from any type of protein–protein interaction screen. The signature descriptor is extended to protein pairs by taking protein signature products. The signature product is implemented as a kernel function within a support vector machine (SVM) classifier.

The importance of understanding protein–protein interactions has prompted the development of various experimental methods,[15] which include two-hybrid screens,[16] mass spectrometry,[17] and protein chips.[18] At the same time, computational techniques have been developed to determine protein–protein interactions based on genomic sequence analysis.[19] These techniques generally employ algorithms to search sequence information for patterns that might be expected to occur based on a priori biological knowledge. Different algorithms search for the conservation of gene neighborhoods and gene order,[20] the occurrence of fused genes,[21,22] the co-evolution of interacting proteins,[23] and the similarity of phylogenetic trees.[24,25] A review of these and other approaches can be found elsewhere.[19,26]

An alternative approach is to use experimental data rather than a priori expectations to guide the discovery of inherent patterns in the sequence data. The method of Sprinzak and Margalit[27] characterizes proteins using known InterPro structural domains, and then uses experimental binding data to identify structural domain pairings correlated with protein–protein binding. This approach uses the inherent assumption that all binding interactions occur within these well-defined domain–domain interactions. Bock and Gough[28,29] take a more general approach by creating protein structural and physiochemical descriptors based on the primary sequence information, and then training an SVM learning system to identify protein–protein binding interactions from these descriptors.

SVMs are classifiers that have been applied successfully to a wide variety of problems in biology as reviewed by Noble.[30] When using an SVM it is necessary to construct a kernel function. A kernel function measures the similarity between two inputs as an inner product. The kernel first used for predicting protein–protein interactions[28] represents proteins as feature vectors based on the physical properties of amino acids. We introduce a kernel function based on signature.[14] This kernel is compared with kernels based on feature vectors and other techniques.

The Signature Kernel

As seen in Section 2, signature can be used to obtain vector representations for chemicals. Trivially, one can derive vector representations for proteins by replacing atoms with amino acids in the signature definition [equation (2)]. In other words, given a protein P, the height h signature is defined as

$$^{h}\sigma(P) = \sum_{a \in P} {}^{h}\sigma_{P}(a) = \alpha(P, {}^{h}S) \tag{6}$$

where ^{h}S is the alphabet of all possible amino acid signatures of height h and $\alpha(P, {}^{h}S)$ is the vector of occurrence numbers of amino acid h-signatures for protein P. Such a signature is computed over the 20 amino acid names/symbols rather than over the elements of the periodic table. As an example, consider the six-letter amino acid sequence *LVMTTM*. All height-1 signatures are based on trimers, and there are four trimers in this sequence: *LVM, VMT, MTT*, and *TTM*. Each signature consists of a root (the middle letter) and its two neighbors, ordered alphabetically. Thus, the signatures corresponding to the four trimers are $V(LM)$, $M(TV)$, $T(MT)$, and $T(MT)$, so that $^{1}\sigma(LVMTTM) = V(LM) + M(TV) + 2T(MT)$. Notice that *MTT* and *TTM* generate the same signature (due to symmetry) and therefore contribute two occurrences to the sum.

We can define an SVM kernel function using equation (6). Since this kernel function is based on the use of the signature descriptor, it is called the *signature kernel*. The signature kernel between two proteins A and B is defined as

$$\begin{aligned} &^{h}k : \{^{h}S\}^{2} \rightarrow \mathbb{R} \\ &^{h}k(A, B) = \alpha(A, {}^{h}S) \cdot \alpha(B, {}^{h}S) \end{aligned} \tag{7}$$

It is worth noting that the signature kernel is very similar to the string kernels of Leslie et al.[31] The only real difference is that signature produces a symmetric formulation, while the strings of Leslie et al. do not. The signature kernel combines signature with SVMs and thus inherits all of the advantages of the two methods. It also allows the use of a very large number of signatures (up to 100,000 in the applications considered later). This would not be feasible with other methods, such as multiple linear regression, that was used in Section 2.

The Signature Product Kernel

As presented above, the signature kernel will only work with data points that consist of a single amino acid sequence. This is a problem since the data points are, in fact, pairs of amino acid sequences (protein–protein pairs). To overcome this difficulty, signature must be defined for pairs of amino acid sequences, and a kernel giving the inner product between two protein–protein pairs must also be derived. To define signature for protein–protein pairs, the notion of tensor product between vectors is used. The tensor product definition can be found in many standard texts on linear algebra. In the present case, the tensor product between $\mathbf{a} = (a_1, \ldots, a_n)^T \in \mathbb{R}^n$ and $\mathbf{b} = (b_1, \ldots, b_n)^T \in \mathbb{R}^m$ is defined as $\mathbf{a} \otimes \mathbf{b} = (a_1 b_1, a_1 b_2, \ldots, a_1 b_m, a_2 b_1, \ldots, a_n b_m)^T \in \mathbb{R}^{nm}$. Using this definition, the signature for pairs, or the signature product, is

$$ {}^h\sigma(A, B) = {}^h\sigma(A) \otimes {}^h\sigma(B) = \alpha(A, {}^hS) \otimes \alpha(B, {}^hS) \tag{8} $$

Using equation (8), an SVM can be applied by specifying a kernel ${}^hk$ that gives a dot product in the signature product space. Using the standard Euclidean inner product the signature product kernel is defined by

$$ {}^hk((A, B), (C, D)) = (\alpha(A, {}^hS) \otimes \alpha(B, {}^hS)) \cdot (\alpha(C, {}^hS) \otimes \alpha(D, {}^hS)) \tag{9} $$

where $\alpha(X, {}^hS)$ is the vector of occurrence numbers of amino acid signatures for protein X. Martin et al.[14] have shown that

$$ \begin{aligned} &(\alpha(A, {}^hS) \otimes \alpha(B, {}^hS)) \cdot (\alpha(C, {}^hS) \otimes \alpha(D, {}^hS)) \\ &\quad = [\alpha(A, {}^hS) \cdot \alpha(C, {}^hS)] \cdot [\alpha(B, {}^hS) \cdot \alpha(D, {}^hS)] \end{aligned} \tag{10} $$

Consequently,

$$ {}^hk((A, B), (C, D)) = {}^hk(A, C) \quad {}^hk(B, D) \tag{11} $$

Equation (11) shows that the signature product kernel of two protein–protein pairs can be computed from the signature kernels of the proteins involved.

Additional improvements in the signature product can be obtained by enforcing symmetry in the protein–protein order. In other words, a protein pair (A, B) can be made equivalent to a protein pair (B, A). This symmetry is easily achieved by defining the symmetric signature product ${}^h\sigma_{\mathrm{SYM}}(A, B) = {}^h\sigma(A) \otimes {}^h\sigma(B) + {}^h\sigma(B) \otimes {}^h\sigma(A)$. The associated symmetric signature product kernel is then

$$ {}^hk_{\mathrm{SYM}}((A, B), (C, D)) = 2({}^hk(A, C)\,{}^hk(B, D) + {}^hk(A, D)\,{}^hk(B, C)) \tag{12} $$

Another improvement is to make use of a normalized dot product to compensate for potential differences in the length of the amino acid sequences involved in the calculations. In particular, a normalized version of the signature kernel (or any kernel) can be implemented as

$$^{h}k_{\mathrm{NORM}} = \frac{^{h}k(A, B)}{\sqrt{^{h}k(A, A)^{h}k(B, B)}} \tag{13}$$

This kernel extends directly to the signature product kernel and is only slightly more complicated in the case of the symmetric signature product.

Predicting Protein–Protein Interactions

We test the signature product kernel in this section with publicly available data sets for *Saccharomyces cerevisiae* (yeast)[32,33] and *Helicobacter pylori*.[34]

Yeast Data Sets The yeast phage display data set consists of 20 yeast SH3 domains and various associated ligands determined through phage display to bind (or not) to the 20 domains, resulting in a data set of 709 domain–ligand pairs. These domain–ligand pairs were processed by calculating different position-specific scoring matrices (PSSMs) to produce an interaction prediction. PSSMs are computed following the procedure described by Tong et al.,[32] resulting in one PSSM per domain.

The results are shown in Table 4. In this table, the signature product is compared to the ligand-only signature as well as to the PSSM method. Tenfold cross-validation is used to calculate accuracy, precision, and sensitivity. Specifically, the data set is divided randomly into 10 equally sized subsets. Each subset is used in turn as a test set, while the method is trained on the union of the remaining nine subsets. The performance of the classifier is evaluated by computing accuracy: $(TP + TN)/(TP + FP + TN + FN)$, precision: $TP/(TP + FP)$, and sensitivity: $TP/(TP + FN)$, where TP, TN, FP, and FN are true or false positive or negative predictions. The values reported in Table 1 are the percentage values of the accuracy, precision, and sensitivities averaged over the 10 test sets.

Another yeast data set extracted from the Database of Interacting Proteins (DIP)[33] is used to test signature on a larger, more heterogeneous data set, and to compare signature performances with the work already done by Sprinzak and Margalit.[27] As in the yeast phage display data set, 10-fold cross-validation is utilized to assess the methods. In contrast to phage display, noninteracting protein pairs were missing in this data set. As also done by Bock and Gough,[29] it is assumed that the pairs not specified explicitly as interacting are noninteracting. From these noninteracting pairs, a set of negative examples are selected at random. It is worth

TABLE 4 Comparison of Signature Kernels with Other Methods[a]

	Accuracy	Precision	Sensitivity
Yeast phage display			
Ligand signature only	73.7	75.5	63.1
Signature product	80.7	81.4	75.2
PSSM	75.4	68.8	81.3
Yeast DIP			
Signature product	69.0	71.5	63.2
Signature InterPro	70.8	86.5	49.2
Sprinzak	68.8	79.8	50.0
H. pylori			
Signature product	83.4	85.7	79.9
Bock and Gough	75.8	80.2	69.8

[a]The signature product method is compared with alternatives using yeast and *H. pylori* data sets. "Ligand signature only" refers to using signature without the signature product method; PSSM is the method used in Tong et al.;[32] signature InterPro uses the signature product but with InterPro entries[35] instead of sequence-based signatures; "Sprinzak" refers to the method in Sprinzak and Margalit,[27] which also uses the InterPro entries; and "Bock and Gough" refers to results reported on *H. pylori*.[29]

noting here (data not shown) that the overall results that are obtained are not affected by different random sets of negative pairs. To the starting number of 2082 interacting protein pairs, another 2082 noninteracting pairs were added, producing a data set with a total of 4164 protein pairs. For the DIP data set we used three methods: the standard signature product, the product kernel with InterPro entries instead of signatures, and the method described by Sprinzak and Margalit.[27] These methods are shown in Table 4 as Product, InterPro, and Sprinzak, respectively.

H. pylori *Data Set* The signature product kernel is used here with two-hybrid measurements of *H. pylori* produced by Rain et al.[34] This data set allows us to examine an organism besides yeast and gives a comparison of the signature method with the method of Bock and Gough.[28,29] Tenfold cross-validation is used to assess the method, and it is assumed that the pairs not specified explicitly as interacting were noninteracting. From these noninteracting pairs, negative examples were selected at random, resulting in 1458 positive and 1458 negative protein pairs. The results of the cross-validation study are shown in Table 4.

Protein-Binding-Domain Prediction

The signature product kernel makes use of signature pairs, which can be viewed as subsequence pairs. The use of subsequence pairs is similar to the correlated InterPro entry pairs described by Sprinzak and Margalit.[27] However, instead of using InterPro entries,[35] signatures are generated automatically and depend only on sequence information. To further establish the relationship between signature and the InterPro domain used by Sprinzak and Margalit,[27] the potential of signature for predicting domains was tested. For this exercise, the yeast DIP data set was used where the top 10 pairs predicted to interact by the signature product kernel were selected. Using the 14 proteins present in these pairs, domain-sized amino acid subsequences were constructed by sliding a window across each of the protein sequences. The window size was 50 and the window was moved in increments of 10 amino acid residues. This method resulted in 1681 subsequences, each 50 amino acids long. The model derived from the yeast DIP data set was then used to predict which pairs of these subsequences would interact. By examining the positions of these interacting subsequences within the full protein sequences, domain predictions could be made as shown in Figure 2.

Figure 2 illustrates domain predictions for P09547 and P50875. In particular, Figure 2(a) shows that in P09547 the region between 300 and 400 is more likely to bind with other regions (windows) in the 14 proteins examined. This region may be a hypothetical domain. Figure 2(b) shows that this domain binds with itself and therefore that P09547 binds with itself. This is only a prediction, but the results are similar to a known interaction, that of P09547 with P50875. It can be seen in Figure 2(c) that P09547 binds with P50875 and that the previously hypothesized domain

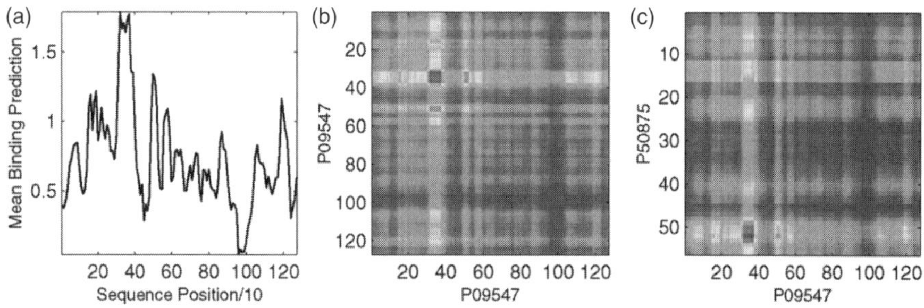

FIGURE 2 Protein-domain predictions. Plots showing the domain predictions for P09547. The x-axis of each plot gives the position of a 50-residue window moved 10 residues at a time across the full sequence of P09547. Plot (a) shows the mean binding activity of the windows along with x-axis with the other 1681 windows considered in the domain prediction example; plot (b) shows an intensity plot of the binding activities of all pairs of windows in P09547; and plot (c) shows an intensity plot of the binding activities of all pairs of windows in P09547 and P50875 (a known binder). In (b and c), red denotes activity and blue denotes inactivity. (*See insert for color representation of the figure*.)

binds to regions 100 to 150, 200 to 250, 400 to 450, and 500 to 550. Again, these regions may be hypothetical domains, this time in P50875. To test whether these predictions match known information, domain information for P09547 and P50875 was searched for in the Swiss-Prot database. The search resulted in two domains mentioned for P09547: an Asn/Thr-rich region at 5–65, and a Gln-rich region at 337–385. The hypothetical domain correlates well with the Gln-rich region. For P50875, Swiss-Prot gave five domains: a poly-Gln domain at 157–162, a poly-Ser domain at 235–240, another poly-Ser domain at 422–425, a poly-Ala domain at 454–463, and a poly-Asn domain at 552–559. Although not perfectly, these domains also correlate with the hypothetical domains found in P50875. Finally, it is worth noting that the domain predictions for P09547 matched with the InterPro entry IPR001660 SAM domain.

The signature method is most similar to the methods of Bock and Gough[28] and Sprinzak and Margalit.[27] Both the signature method and that of Bock and Gough use SVMs, sequence information, and experimental data to predict protein–protein interaction. However, while Bock and Gough transform sequence information into physicochemical information (charge, hydrophobicity, and surface tension), the signature descriptor does not require such a transformation. That signature performs as well as physicochemical properties may be due to the fact that physicochemical properties can be derived from signature through QSPR (see Section 2).

Although the signature product kernel has been benchmarked fairly extensively and appears to perform well, some issues and caveats need to be outlined. First, the signature method requires some type of input, either from experiment or prediction. If the data are unreliable, as is often the case with protein–protein interaction data,[36,37] the results will also be unreliable. Second, the fact that there are so few interacting protein pairs relative to the number of possible pairs can cause problems. In particular, there are bound to be many false-positive predictions even with a good classifier. Although in this regard the signature method performs as well as the method described by Jansen et al.,[38] it nevertheless remains a practical problem (for any classifier) given the naturally unbalanced population of protein–protein interactions.

In sum, signature can be used to construct kernels for predicting protein–protein interactions. The method uses a unique product description of protein pairs called *signature product*. Motivated by the results presented in this section, in the next section the signature product is extended to both proteins and chemicals to facilitate detection of enzyme–metabolite and drug–target interactions.

4 PROTEIN–CHEMICAL SIGNATURE: PREDICTING ENZYME–METABOLITE AND DRUG–TARGET INTERACTIONS

Databases today contain sufficient information linking biological (protein sequences) and chemicals (metabolites and drugs) to apply machine learning methods for predicting interactions among chemicals and proteins. Machine learning techniques have generally used as input either protein sequence or chemical structure information. The approach described in this section can infer protein–chemical interactions using heterogeneous input consisting of both sequence and chemical information.[39] The approach is illustrated using data from the KEGG database[40] linking metabolites to enzymes, and data from the DRUGBANK database[41] linking drugs to targets. The technique is benchmarked using cross-validation to predict when a given enzyme will catalyze a given metabolite, or when a given drug will bind to a given target. The generalization ability of the approach is demonstrated by making predictions on enzyme–metabolite and drug–target pairs in the absence of prior binding information about an enzyme, a metabolite, a drug, or a target.

Enzyme–Metabolite and Drug–Target Interactions

Molecular recognition is the primary event involved in the interaction of proteins with other proteins and with small molecules such as metabolites and therapeutics. The ability to predict these interactions is vital to answering the following questions on a genome-wide scale: *Which enzymes can catalyze a given metabolic reaction? Which reactions can be catalyzed by a given enzyme? To which targets can a given drug bind? Which drugs will bind to a given target?* The rapid increase in the amount of chemical information and number of protein sequences being deposited to databases and the linking of chemical information to biological information in several data repositories[40–44] has provided the required data sets on which machine learning tools can be applied to these questions.

Molecular recognition has been studied for quite some time using a variety of approaches. At the molecular level, biophysical simulation techniques ranging from ab initio quantum chemical calculations to rapid virtual screening[45,46] have been used. However, these techniques require knowledge of the 3D structure of the chemical–protein interface and can be computationally expensive, giving them limited applicability at the whole genome level. The advantages of machine learning approaches are their computational efficiency and the fact that they do not necessarily require 3D structures but can perform using only biological sequences and chemical molecular connectivity.

To date, machine learning methods have been applied to molecular recognition questions based on whether the input is biological sequences or chemical connectivity data. In bioinformatics, for example, there are techniques making use of the Enzyme Commission (EC) numbers to predict the metabolites a given sequence can catalyze.[47–49] There are also sequence- and structure-based methods for locating homologous ligand-binding sites.[50,51] In cheminformatics, methods have been developed to find compounds binding to a given target[52] as well as to predict the EC number of a given metabolic reaction.[53] However, none of these techniques take advantage of the increasing linkage between biological sequences and molecular information. Further, there are no methods available that can predict when an unclassified (in the EC nomenclature) chemical reaction will be catalyzed by a given enzyme, or when a chemical will bind to a sequence in the absence of binding information for either the sequence or the chemical. In this section, a unified approach to these problems is proposed using an SVM[30] kernel-based method that operates on databases linking both chemical and biological information.

The approach outlined here uses the signature descriptor and, in this case, both proteins and chemicals are encoded with *atomistic* signatures. Signature and signature-like descriptors such as MOLPRINT-2D[54] have been used to measure similarity between chemicals. Predictions using proteins are also based on measuring similarity between proteins sequences. These "similarity" measurements are typically made using homology and are computed by sequence alignment algorithms, or, in the case of SVMs, by remote homology detection with string kernels.[31] Protein sequence–based approaches are used to predict interaction, function, and subcellular localization, and the tool routinely used to assign EC numbers in newly sequenced genome is sequence alignment.[55]

The similarities between the chemical- and sequence-based approaches described above provided the motivation for combining these methods into a general framework. To accomplish this goal, molecular signature is used as a common representation for both molecules and protein sequences. Molecules cannot be coded as linear sequences of well-defined units, but protein sequences are chemicals and as such can be represented by molecular graphs. A common representation should thus come from cheminformatics. The choice of molecular signature (among the myriad cheminformatics descriptors) is motivated by the fact that signatures are molecular fragments. When applied to proteins, these fragments correspond to short sequences, and short sequences correspond to protein domains, or DNA motifs, both of which are used routinely in bioinformatics applications.

The approach is implemented within the framework of SVMs by developing kernels that use the signature descriptor in the context of protein–chemical pairs. The kernels encompass ideas related to string kernels,[31] graph kernels,[56–59] and pairwise kernels.[14,60]

Signature Kernels and Signature Product Kernels

Several kernels are needed, depending on the object manipulated: chemical, reaction, or sequence. When the objects of interest are chemicals, the kernel function computes a similarity between two chemicals represented by their molecular signatures. In this chapter, two similarity measures are used: a simple dot product between the molecular signature vectors and a Tanimoto similarity index as introduced by Swamidass et al.[59] When the objects of interest are chemical reactions, the kernel function then computes a similarity between the signatures of the reactions. When dealing with protein sequences, the sequences are first transformed into molecular graphs by replacing the amino acids with their chemical structures. The proteins are then processed as chemicals. Finally, when using both chemicals and proteins, as in the case of a drug binding to a target or a metabolic reaction catalyzed by an enzyme, the signature product kernel[14] is used with atomistic signatures. As mentioned in Section 3, this kernel was developed originally for protein sequences and applied to predict protein–protein interactions. In the present situation, the kernel is defined for two pairs of protein-chemical interactions. It is the product of the signature kernel for the two chemicals and the signature kernel for the two proteins. Equations (14) to (16) formally define these kernels.

Using the same notations as in the earlier sections, for a given height h and a given alphabet ${}^{h}\Sigma$ of all atomic signatures of height h, the signature kernel between two chemical structures A and B is defined by

$$ {}^{h}k(A, B) = \frac{\alpha(A, {}^{h}\Sigma) \cdot \alpha(B, {}^{h}\Sigma)}{\|\alpha(A, {}^{h}\Sigma)\| \cdot \|\alpha(B, {}^{h}\Sigma)\|} \tag{14} $$

where $\alpha(A, {}^{h}\Sigma)$ and $\alpha(B, {}^{h}\Sigma)$ are computed using equation (2), and $||\alpha(A, {}^{h}\Sigma)||$ denotes the norm of $\alpha(A, {}^{h}\Sigma)$. This kernel also applies to proteins considered as atomic structures and reactions, where reaction signatures are used instead of molecular signatures.

The Tanimoto signature kernel is defined as

$$ {}^{h}k_t(A, B) = \frac{{}^{h}k(A, B)}{{}^{h}k(A, A) + {}^{h}k(B, B) - {}^{h}k(A, B)} \tag{15} $$

where ${}^{h}k(A, B)$ is the signature kernel of height h as defined in equation (14). Due to restriction on the Tanimoto kernel, a binary-only representation is used for $\alpha(A, {}^{h}\Sigma)$ and $\alpha(B, {}^{h}\Sigma)$. In this case all nonzero components of the molecular signature vectors are replaced by the value 1.

Let (C_1, P_1) and (C_2, P_2) be two pairs of chemical–protein interactions, the signature product kernel is defined as

$$
{}^{h,l}k_p((C_1, P_1), (C_2, P_2)) = {}^{h}k(C_1, C_2)\,{}^{l}k(P_1, P_2) \qquad (16)
$$

where ${}^{h}k(C_1, C_2)$ is the signature kernel of height h for the chemical pair (C_1, C_2), and ${}^{l}k(P_1, P_2)$ is the signature kernel of height l for the protein pair (P_1, P_2), and both kernels are defined using equation (14).

Classifying Compounds with Signature

Before predicting protein–chemical interactions, the signature kernel was tested and compared with state-of-the-art kernels for the classification of chemicals using the Predictive Toxicology Challenge (PTC) data set.[61] This data set provides the carcinogenicity of 417 chemical compounds for male mice (MM), female mice (FM), male rats (MR), and female rats (FR). Although toxicology is generally difficult to predict, this data set has been used previously to benchmark various kernel methods.[59] Table 5 gives a comparison of the Tanimoto signature kernel [equation (15)] with a frequent pattern discovery approach,[57] a marginalized kernel,[62] SMILES string kernels, path count–based graph kernels, and a kernel based on 3D atomic distances.[59] The Tanimoto signature kernel compares well with these state-of-the-art kernels. The fact that the signature kernel outperforms methods based on 3D chemical structures is consistent with previous structure–activity analyses, where signature was compared with various one-, two-, and three-dimensional molecular descriptors.[7]

Predicting EC Numbers

Enzymes and metabolic reactions are organized according to the EC system, a hierarchical classification that assigns unique four-field numbers to different enzymatic activities.[63] The first field of an EC number indicates the general class of catalyzed reaction: 1, oxidoreductases; 2, transferases; 3, hydrolases; 4, lyases; 5, isomerases; and 6, ligases. The second and third fields depend on different criteria related to the chemical features of the substrate and the product of the reaction, and the fourth

TABLE 5 Leave-One-Out Accuracies (%) for Chemical Toxicity Prediction[a]

Method	MM 336 (38.4)[b]	FM 349 (41.0)	MR 344 (44.2)	FR 351 (34.5)
Pattern discovery[c]	61.0	61.0	62.8	66.7
Marginalized kernel[c]	64.3	63.4	58.4	66.1
SMILES string kernel[d]	66.4	63.0	57.6	67.0
Tanimoto 2D path kernel[d]	66.4	64.5	**65.7**	66.9
3D kernel[d]	59.8	61.0	60.8	64.4
Tanimoto Signature kernel $h = 0$	66.1	60.2	59.3	68.7
$h = 0–1$[e]	65.5	60.2	61.0	**69.2**
$h = 0–2$	65.8	63.0	63.4	68.1
$h = 0–3$	**66.7**	64.8	63.7	67.8
$h = 0–4$	66.1	**66.5**	63.7	67.2
$h = 0–5$	64.9	65.9	63.7	67.5
$h = 0–6$	64.3	65.9	62.8	67.5

[a]Toxicity is predicted for various species and genders. MM: male mice, FM: female mice, MR: male rats, FR: female rats

[b]Denotes the data set size (e.g., 336) and the percentage of positives in the data set (e.g., 38.4%).

[c]From Kashima et al.[57]

[d]From Swamidass et al.[59]

[e]The Tanimoto signature kernel is computed for all signatures with height in the range of, for example, 0–1.

field is substrate and product specific. For example, tripeptide aminopeptidase is assigned the number EC 3.4.11.4. Level 1: "EC 3" enzymes are hydrolases. Level 2: "EC 3.4" are hydrolases that act on peptide bonds. Level 3: "EC 3.4.11" are those hydrolases that cleave off the amino-terminal amino acid from a polypeptide. Level 4: "EC 3.4.11.4" are those that cleave off the amino-terminal end from a tripeptide.

Using Metabolic Reactions To predict the EC numbers of metabolic reactions, a training set was created by downloading the KEGG database. At the time of these calculations, this database contained 10,951 compounds and 6556 reactions involving compounds with molecular structures stored in the database. The signature reaction kernel [equation (14)] was applied in order to classify metabolic reactions. Cross-validation was performed for the first three EC levels, with results presented in Figure 3(a). The average accuracies are above 80%, which is lower than those published in Kotera et al.[53] However, the signature kernel can process a larger number of reactions.

Using Protein Sequences Protein sequences having one or more EC number assignments were downloaded from the KEGG database, resulting

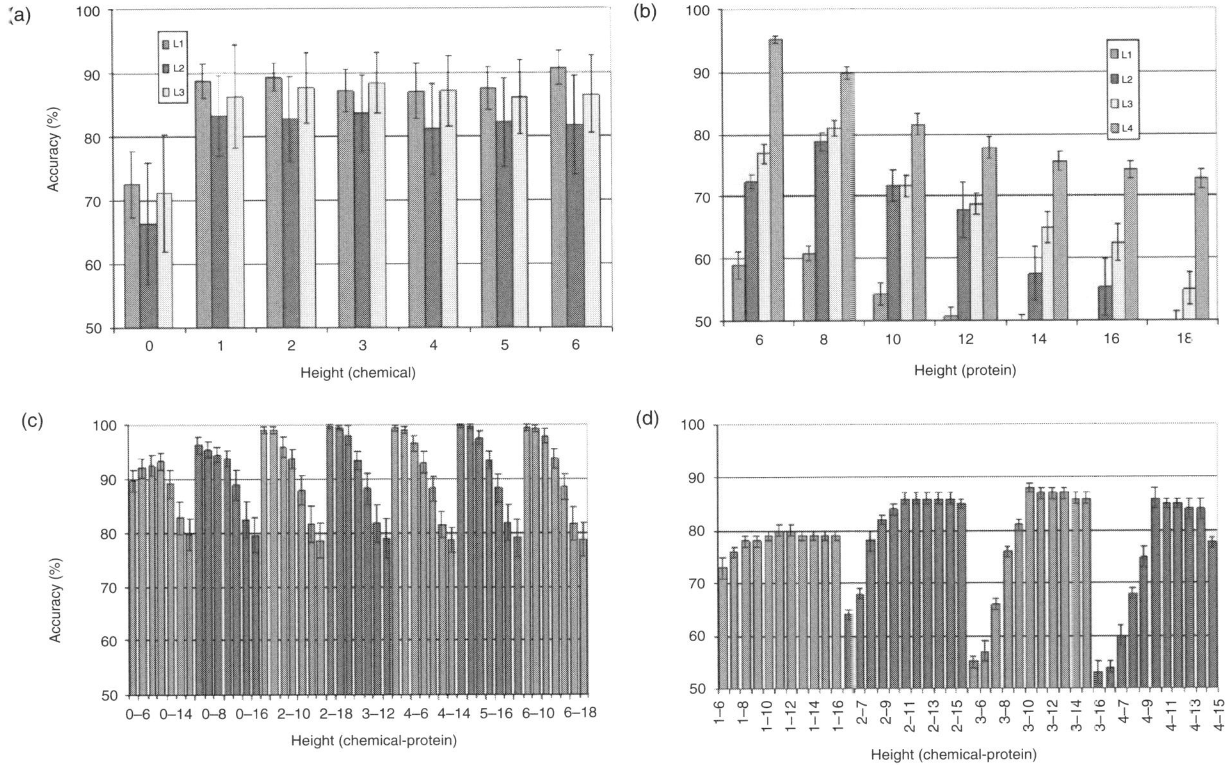

FIGURE 3 Fivefold cross-validation prediction accuracies for EC numbers. Signature kernels for various chemical, proteins, and signature heights are used. (a) Accuracies of the signature kernel for the first three EC levels (L1 to L3) using only chemical reactions. (b) Accuracies of the signature kernel for the four EC levels (L1 to L4) using only protein sequences. (b) Accuracies of the signature product kernel for the EC class 1.1.1.1. (d) Accuracies for the KEGG drug–target data set using the signature product kernel. (*See insert for color representation of the figure*.)

in 308,094 sequences. The signature kernel was computed for each protein sequence using an atomistic description of the protein. Cross-validation was performed using all EC levels and various signature heights. Results are presented in Figure 3(b). These accuracies are generally lower than those reported using protein structure graph kernels,[47] feature vectors (i.e., SVM-PROT),[48] or motifs.[49] Nevertheless, the accuracies are high enough so that the signature kernel can be used with the protein sequence alone to predict EC numbers (except for EC level 1); with sequence *and* reaction, the accuracy improves.

Using Both Metabolic Reactions and Protein Sequences Using the reactions and sequences with assigned EC numbers in the KEGG database, a set of 855,772 pairs was compiled (out of 3905 reactions and 255,304 enzymes). The product signature kernel [equation (16)] was applied for each EC level using various chemical and protein signature heights. Cross-validation results are reported for class 1.1.1.1 in Figure 3(c). Additional cross-validation results and other comparisons are summarized in Table 6. Table 6 shows that the signature product outperforms all other techniques in terms of sensitivity (accuracy on positives) and Jaccard coefficient. A higher sensitivity and Jaccard coefficient (J), with $J = TP/(TP + FP + FN)$, indicates lower numbers of false positives and negatives. Consequently, the signature product kernel can be used to accurately process a larger number of sequences than can be processed correctly with the other techniques listed in Table 6.

Predicting Novel Enzyme–Metabolite Interactions To test the ability of the signature product kernel to predict unknown enzyme–metabolite interactions, a list of enzymes and reactions corresponding to EC numbers accepted in September 2006 by the Nomenclature Committee of the International Union of Biochemistry and Molecular Biology (NC-IUBMB) was removed from the KEGG database. This list was used as a test set to assess the ability of the signature product kernel to predict when an enzyme will catalyze a reaction where neither the reaction nor the enzyme is characterized by an EC number. A training set from the reduced KEGG database was compiled. The results are similar to those of signature product L1 in Table 6 and demonstrate that it is possible to predict (with accuracies above 80%) whether or not a given enzyme will catalyze a given reaction, even when the EC class of the enzyme reaction pair is not present in the training set.

TABLE 6 Comparison of Signature Product Kernels with Other SVM Techniques[a]

Level	Method	Accuracy	AuC	Precision	Sensitivity	Specificity	J
L1	Graph kernel[b]	89.9 ± 1.3			40.0 ± 7.4	99.9 ± 0.1	
	Signature product	88.0 ± 6.1	91.8 ± 4.6	87.1 ± 6.8	89.6 ± 6.1	86.4 ± 7.2	79.4 ± 9.6
L2	SVM-PROT[c]	95.2 ± 5.4		97.4 ± 3.8	77.4 ± 14.1	88.6 ± 7.6	70.6 ± 13.8
	Signature product	94.2 ± 4.9	96.0 ± 4.0	93.6 ± 5.7	95.2 ± 4.6	93.3 ± 6.3	89.6 ± 8.1
L3	MEX-motifs[d]						89.3 ± 8.0
	Signature product	97.9 ± 2.3	98.9 ± 1.2	97.9 ± 3.1	97.9 ± 2.1	97.9 ± 3.4	96.0 ± 4.1
L4	Signature product	99.0 ± 2.3	99.2 ± 2.0	98.7 ± 2.7	95.5 ± 2.9	98.7 ± 2.7	98.2 ± 4.1

[a]EC numbers are predicted at all hierarchical levels. For the signature product kernel, statistics were calculated using fivefold cross-validation. Accuracy, precision, sensitivity, and specificity are defined in the text. The table also reports the Jaccard coefficient (J) and the area under the receiver operator characteristic (ROC) curve. Briefly, the area under the curve (AuC) is obtained by plotting the ROC curve (i.e., TP rate = specificity vs. FP rate = 1-sensitivity for varying thresholds separating positives from negatives). For all these statistics, a larger number indicates a better result.
[b]From Borgwardt et al. [47]
[c]From Cai et el. [48]
[d]From Kunik et al. [49]

Predicting Drug–Target Binding

A data set linking drugs with protein targets was created from the KEGG database. This data set contained 873 drug–target pairs taken from 121 targets and 551 drugs. Cross-validation results were obtained using five-fold cross-validation with the signature product kernel, as shown in Figure 3(d). The same training set was used to predict if any of the drug target binding pairs stored in the DRUGBANK database could be predicted using the SVM trained on the KEGG database. At the time of these calculations, DRUGBANK contained 1133 drugs and 509 targets, with 1849 drug–protein pairs. Out of the 1133 drugs, 124 had a name and a structure stored in KEGG. Consequently, out of the 1849 pairs, only 298 pairs could potentially be predicted from the KEGG training set. There were only 32 pairs in common between KEGG and DRUGBANK. Despite this small number, 189 additional interactions not in KEGG predicted by the signature product kernel were found in DRUGBANK, including 67 interactions between drugs and targets not present in KEGG (Figure 4).

The fact that the signature kernel can predict when a new drug will bind to a given target when other binders are known for that target is not unexpected. Indeed, it has already been demonstrated that signature and signature-like descriptors can be used to derive structure-activity relationships, and thus predict other drugs binding to a specific target.[7,64] The more interesting result is that signature and signature product kernels can be used to accurately predict drugs binding to a target even in the absence

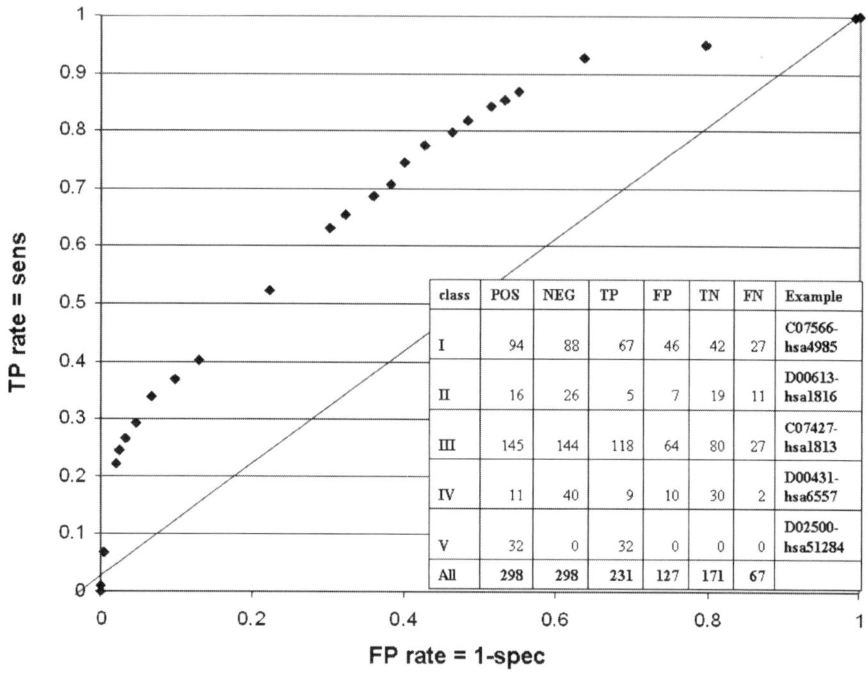

class	POS	NEG	TP	FP	TN	FN	Example
I	94	88	67	46	42	27	C07566-hsa4985
II	16	26	5	7	19	11	D00613-hsa1816
III	145	144	118	64	80	27	C07427-hsa1813
IV	11	40	9	10	30	2	D00431-hsa6557
V	32	0	32	0	0	0	D02500-hsa51284
All	**298**	**298**	**231**	**127**	**171**	**67**	

FIGURE 4 ROC curve for predicted DRUGBANK interactions not present in KEGG. The KEGG-derived training set contained 873 positive drug–target pairs and 873 randomly generated negative (not in KEGG) pairs. The DRUGBANK-derived test set contained 298 positive interactions and 298 randomly generated negative (not in DRUGBANK) pairs. A product signature SVM was obtained using the training set with height-2 for drugs and height-10 for proteins. To avoid training set bias, this procedure was repeated 100 times for 100 different choices of negative pairs. An averaged SVM prediction value was computed for each element of the test set. The ROC curve in this figure was generated by varying the threshold for calling a test pair positive or negative. The area under the ROC curve is 0.74. The table reports prediction results for an FP rate of 0.42. This rate maximizes the overall accuracy (67.5%). The results are divided into five possible cases, depending on whether or not the targets or the drugs from the DRUGBANK test set were present in the KEGG training set: class I contains cases where neither the target nor the drug is in the training set; class II contains cases where the drugs are in the training set but the targets are absent; class III contains cases where the targets are in the training set but not the drugs; class IV contains cases where both the drug and the target are in the training set, albeit with different partners; and class V is composed of the 32 interactions common between KEGG and DRUGBANK. Examples of predicted interactions are listed in the last column using the KEGG nomenclature for targets and drugs.

of prior binding information for the target (or when target or drug are not present in the training set).

Implications of Signature-Based Kernels for Metabolic Network Inferences

The method proposed for predicting protein–chemical interactions is based on the representation of a protein using atomistic structure. A comparison with sequence alignment scores reveals that signature could potentially be used to detect protein homology,[39] even though (strictly speaking) sequence information is ignored in favor of molecular structure. This is perhaps surprising. Looking at the atomistic structures of the amino acids, it can be demonstrated that a height-2 atomic signature can characterize each amino acid uniquely, every dimer can be characterized using a height-8 atomic signature, every trimer can be characterized with a height-10 signature, and so on. Thus, when two proteins have many signatures in common, they share many amino acid strings and thus should have some degree of homology. Sharing strings certainly increases alignment score, but there are some instances where strings made of different amino acids can lead to high homology scores. Sequence alignment algorithms incorporate this information in substitution matrices such as BLOSUM62. The surprising implication of our comparison with the sequence alignment score is that an *atomistic structure representation of proteins encompasses information stored in substitution matrices*. This is the key observation that underlies the method.

An interesting consequence of this observation is that the signature kernel and the product kernel can be used to classify proteins into the EC nomenclature. EC numbers are a hierarchical classification of enzymatic reactions, commonly used to annotate proteins while analyzing genomes. EC numbers therefore link gene sequences to reactions. Techniques have been proposed to predict EC numbers using either chemical information or enzyme sequences. The techniques based on chemical information[53] can predict the hierarchical category of a given reaction, with the exception of the fourth level, which is reaction specific. The sequence-based techniques[47–49] can predict the EC numbers at all levels for a given sequence, although these methods perform poorly on the first level. However, none of these techniques can predict when a given protein will catalyze a reaction unless the reaction has a known EC class. Further, the EC class must have a sufficiently large set of associated sequences, such that machine learning can be used. The signature method, on the other hand, can be used to overcome these hurdles.

The traditional method for constructing a metabolic map of a newly sequenced organism is to assign EC numbers to its proteins. Table 6 shows that this task is better performed when both sequence and chemical information are taken into account. Yet many proteins remain unannotated not only because their sequences have not been mapped to an already classified enzyme, but also because the reactions catalyzed by the proteins have not been characterized in the EC nomenclature. EC number assignment requires published evidence and full characterization of the enzymatic reaction. For this reason, many reactions, although occurring in various pathways, do not have an assigned EC number. Addressing this issue, the method proposed here can predict when a protein will catalyze a reaction, even in the absence of any EC nomenclature information.

5 CONCLUSIONS

Cheminformatics and bioinformatics share many common goals. In cheminformatics one may search for similar compounds, whereas in bioinformatics one may look for homologous sequences. In cheminformatics one may predict activities and properties of compounds, whereas in bioinformatics one may want to infer protein function or protein–protein interactions. In cheminformatics one computes the binding affinities between chemicals and proteins with the purpose of designing better drugs. In bioinformatics one predicts the possibility of two proteins to interact, the possibility of a protein to bind to a DNA site, or the possibility of interaction between an enzyme and a metabolite. Beyond common goals, cheminformatics and bioinformatics also share common computational techniques such as clustering and machine learning based on regression, support vector machine (SVM), neural networks, Bayesian networks, hidden Markov models, and decision trees.

Despite these common goals and computational techniques, knowledge exchange between cheminformatics and bioinformatics is limited, and the two fields seem to have grown separately. In this chapter we have attempted to bridge this gap by proposing a common signature representation for both chemicals and sequences. Specifically, we have shown that this common representation performs well for many applications in chemistry and biology. Particularly interesting is the demonstration that signature can be used to answer questions at the chemistry–biology interface, such as predicting protein–chemical interactions. The ability to predict these interactions on a genome-wide scale is vital to determining the enzymes that catalyze a given metabolic reaction, the reactions a given enzyme catalyzes, the targets a particular drug binds, and the drugs that will bind a given target. Predicting these interactions has direct application when completing genome annotations, finding enzymes for synthetic chemistry, predicting drug specificity, promiscuity, polypharmacology, and identifying broad-spectrum therapeutics.

REFERENCES

1. Kier, L. B.; Hall, L. H. Intermolecular accessibility: the meaning of molecular connectivity. *J. Chem. Inf. Comput. Sci*. **2000**, 40, 792–795.
2. Randic, M.; Zupan, J. On interpretation of well-known topological indices. *J. Chem. Inf. Comput. Sci*. **2001**, 41, 550–560.
3. Brown, R. D.; Martin, Y. C. The information content in 2D and 3D structural descriptors relevant to ligand–receptor binding. *J. Chem. Inf. Comput. Sci*. **1997**, 37, 1–9
4. Randić, M.; Basak, S. C. A new descriptor for structure–property and structure–activity correlations. *J. Chem. Inf. Comput. Sci*. **2001**, 41, 650–656.
5. Faulon, J.-L. Stochastic generator of chemical structure: 1. Application to the structure elucidation of large molecules. *J. Chem. Inf. Comput. Sci*. **1994**, 34, 1204–1218.
6. Visco, D. P., Jr.; Pophale, R. S.; Rintoul, M. D.; Faulon, J.-L. Developing a methodology for an inverse quantitative structure–activity relationship using the signature molecular descriptor. *J. Mol. Graph. Model*. **2002**, 20, 429–438.
7. Faulon, J. L.; Visco, D. P., Jr.; Pophale, R. S. The signature molecular descriptor: 1. Using extended valence sequences in QSAR and QSPR studies. *J. Chem. Inf. Comput. Sci*. **2003**, 43(3), 707–720.
8. Faulon, J.-L. Isomorphism, automorphism partitioning, and canonical labeling can be solved in polynomial-time for molecular graphs. *J. Chem. Inf. Comput. Sci*. **1998**, 38, 432–444.
9. Faulon, J. L.; Collins, M. J.; Carr, R. D. The signature molecular descriptor: 4. Canonizing molecules using extended valence sequences. *J. Chem. Inf. Comput. Sci*. **2004**, 44(2), 427–436.
10. Hall, L. H. *MOLCONN-Z*. Hall Associates Consulting, Quincy, MA, 1991.
11. Kier, L. B.; Hall, L. H. *Molecular Structure Description*. Academic Press, San Diego, CA, 1999.
12. Draper, N. R.; Smith, H. *Applied Regression Analysis*, 2nd ed. Wiley, New York, 1981.
13. Tong, W.; Lowis, D. R.; Perkins, R.; Chen, Y.; Welsh, W. J.; Goddette, D. W.; Heritage, T. W.; Sheehan, D. M. Evaluation of quantitative structure–activity relationship methods for large-scale prediction of chemicals binding to the estrogen receptor. *J. Chem. Inf. Comput. Sci*. **1998**, 38(4), 669–677.

14. Martin, S.; Roe, D.; Faulon, J. L. Predicting protein-protein interactions using signature products. *Bioinformatics* **2005**, 21(2), 218–226.
15. Shoemaker, B. A.; Panchenko, A. R. Deciphering protein–protein interactions: I. Experimental techniques and databases. *PLoS Comput. Biol*. **2007**, 3(3), e42.
16. Fields, S.; Song, O. A novel genetic system to detect protein–protein interactions. *Nature* **1989**, 340(6230), 245–246.
17. Ho, Y.; Gruhler, A.; Heilbut, A.; Bader, G. D.; Moore, L.; Adams, S. L.; Millar, A.; Taylor, P.; Bennett, K.; Boutilier, K.; et al. Systematic identification of protein complexes in *Saccharomyces cerevisiae* by mass spectrometry. *Nature* **2002**, 415(6868), 180–183.
18. Zhu, H.; Bilgin, M.; Bangham, R.; Hall, D.; Casamayor, A.; Bertone, P.; Lan, N.; Jansen, R.; Bidlingmaier, S.; Houfek, T.; et al. Global analysis of protein activities using proteome chips. *Science* **2001**, 293(5537), 2101–2105.
19. Shoemaker, B. A.; Panchenko, A. R. Deciphering protein–protein interactions: II. Computational methods to predict protein and domain interaction partners. *PLoS Comput. Biol*. **2007**, 3(4), e43.
20. Dandekar, T.; Snel, B.; Huynen, M.; Bork, P. Conservation of gene order: a fingerprint of proteins that physically interact. *Trends Biochem. Sci*. **1998**, 23(9), 324–328.
21. Enright, A. J.; Iliopoulos, I.; Kyrpides, N. C.; Ouzounis, C. A. Protein interaction maps for complete genomes based on gene fusion events. *Nature* **1999**, 402(6757), 86–90.
22. Marcotte, E. M.; Pellegrini, M.; Ng, H. L.; Rice, D. W.; Yeates, T. O.; Eisenberg, D. Detecting protein function and protein–protein interactions from genome sequences. *Science* **1999**, 285(5428), 751–753.
23. Pazos, F.; Helmer-Citterich, M.; Ausiello, G.; Valencia, A. Correlated mutations contain information about protein–protein interaction. *J. Mol. Biol*. **1997**, 271(4), 511–523.
24. Goh, C. S.; Bogan, A. A.; Joachimiak, M.; Walther, D.; Cohen, F. E. Coevolution of proteins with their interaction partners. *J. Mol. Biol*. **2000**, 299(2), 283–293.
25. Kumar, A.; Agarwal, S.; Heyman, J. A.; Matson, S.; Heidtman, M.; Piccirillo, S.; Umansky, L.; Drawid, A.; Jansen, R.; Liu, Y.; et al. Subcellular localization of the yeast proteome. *Genes Dev* **2002**, 16(6), 707–719.
26. Valencia, A.; Pazos, F. Computational methods for the prediction of protein interactions. *Curr. Opin. Struct. Biol*. **2002**, 12(3), 368–373.
27. Sprinzak, E.; Margalit, H. Correlated sequence-signatures as markers of protein–protein interaction. *J. Mol. Biol*. **2001**, 311(4), 681–692.
28. Bock, J. R.; Gough, D. A. Predicting protein–protein interactions from primary structure. *Bioinformatics* **2001**, 17(5), 455–460.
29. Bock, J. R.; Gough, D. A. Whole-proteome interaction mining. *Bioinformatics* **2003**, 19(1), 125–134.

30. Noble, W. S. What is a support vector machine ? *Nat. Biotechnol*. **2006**, 24(12), 1565–1567.

31. Leslie, C.; Eskin, E.; Noble, W. S. The spectrum kernel: a string kernel for SVM protein classification. *Pac. Symp. Biocomput*. **2002**, 564–575.

32. Tong, A. H.; Drees, B.; Nardelli, G.; Bader, G. D.; Brannetti, B.; Castagnoli, L.; Evangelista, M.; Ferracuti, S.; Nelson, B.; Paoluzi, S.; et al. A combined experimental and computational strategy to define protein interaction networks for peptide recognition modules. *Science* **2002**, 295(5553), 321–324.

33. Xenarios, I.; Salwinski, L.; Duan, X. J.; Higney, P.; Kim, S. M.; Eisenberg, D. DIP, the database of interacting proteins: a research tool for studying cellular networks of protein interactions. *Nucleic Acids Res*. **2002**, 30(1), 303–305.

34. Rain, J. C.; Selig, L.; De Reuse, H.; Battaglia, V.; Reverdy, C.; Simon, S.; Lenzen, G.; Petel, F.; Wojcik, J.; Schachter, V.; et al. The protein–protein interaction map of *Helicobacter pylori*. *Nature* **2001**, 409(6817), 211–215.

35. Apweiler, R.; Attwood, T. K.; Bairoch, A.; Bateman, A.; Birney, E.; Biswas, M.; Bucher, P.; Cerutti, L.; Corpet, F.; Croning, M. D.; et al. The InterPro database, an integrated documentation resource for protein families, domains and functional sites. *Nucleic Acids Res*. **2001**, 29(1), 37–40.

36. von Mering, C.; Krause, R.; Snel, B.; Cornell, M.; Oliver, S. G.; Fields, S.; Bork, P. Comparative assessment of large-scale data sets of protein–protein interactions. *Nature* **2002**, 417(6887), 399–403.

37. Sprinzak, E.; Sattath, S.; Margalit, H. How reliable are experimental protein–protein interaction data? *J. Mol. Biol*. **2003**, 327(5), 919–923.

38. Jansen, R.; Yu, H.; Greenbaum, D.; Kluger, Y.; Krogan, N. J.; Chung, S.; Emili, A.; Snyder, M.; Greenblatt, J. F.; Gerstein, M. A Bayesian networks approach for predicting protein–protein interactions from genomic data. *Science* **2003**, 302(5644), 449–453.

39. Faulon, J. L.; Misra, M.; Martin, S.; Sale, K.; Sapra, R. Genome scale enzyme–metabolite and drug–target interaction predictions using the signature molecular descriptor. *Bioinformatics* **2008**, 24(2), 225–233.

40. Kanehisa, M.; Goto, S.; Hattori, M.; Aoki-Kinoshita, K. F.; Itoh, M.; Kawashima, S.; Katayama, T.; Araki, M.; Hirakawa, M. From genomics to chemical genomics: new developments in KEGG. *Nucleic Acids Res*. **2006**, 34(Database issue), D354–D357.

41. Wishart, D. S.; Knox, C.; Guo, A. C.; Shrivastava, S.; Hassanali, M.; Stothard, P.; Chang, Z.; Woolsey, J. DrugBank: a comprehensive resource for in silico drug discovery and exploration. *Nucleic Acids Res*. **2006**, 34(Database issue), D668–D672.

42. Austin, C. P.; Brady, L. S.; Insel, T. R.; Collins, F. S. NIH molecular libraries initiative. *Science* **2004**, 306(5699), 1138–1139.

43. Brooksbank, C.; Cameron, G.; Thornton, J. The European Bioinformatics Institute's data resources: towards systems biology. *Nucleic Acids Res*. **2005**, 33(Database issue), D46–D53.
44. Chen, X.; Ji, Z. L.; Chen, Y. Z. TTD: therapeutic target database. *Nucleic Acids Res*. **2002**, 30(1), 412–415.
45. Kellenberger, E.; Rodrigo, J.; Muller, P.; Rognan, D. Comparative evaluation of eight docking tools for docking and virtual screening accuracy. *Proteins* **2004**, 57(2), 225–242.
46. Warren, G. L.; Andrews, C. W.; Capelli, A. M.; Clarke, B.; LaLonde, J.; Lambert, M. H.; Lindvall, M.; Nevins, N.; Semus, S. F.; Senger, S.; et al. A critical assessment of docking programs and scoring functions. *J. Med. Chem*. **2006**, 49(20), 5912–5931.
47. Borgwardt, K. M.; Ong, C. S.; Schonauer, S.; Vishwanathan, S. V.; Smola, A. J.; Kriegel, H. P. Protein function prediction via graph kernels. *Bioinformatics* **2005**, 21(Suppl. 1), i47–i56.
48. Cai, C. Z.; Han, L. Y.; Ji, Z. L.; Chen, X.; Chen, Y. Z. SVM-Prot: Web-based support vector machine software for functional classification of a protein from its primary sequence. *Nucleic Acids Res*. **2003**, 31(13), 3692–3697.
49. Kunik, V.; Solan, Z.; Edelman, S.; Ruppin, E.; Horn, D. Motif extraction and protein classification. *Proc. IEEE Comput. Syst. Bioinf. Conf*. **2005**, 80–85.
50. Johnson, J. M.; Church, G. M. Predicting ligand-binding function in families of bacterial receptors. *Proc. Natl. Acad. Sci. USA* **2000**, 97(8), 3965–3970.
51. Kalinina, O. V.; Novichkov, P. S.; Mironov, A. A.; Gelfand, M. S.; Rakhmaninova, A. B. SDPpred: a tool for prediction of amino acid residues that determine differences in functional specificity of homologous proteins. *Nucleic Acids Res*. **2004**, 32(Web Server issue), W424–W428.
52. Gasteiger, J.; Engel, T. *Chemoinformatics*. Wiley-VCH, Weinheim, Germany, 2003.
53. Kotera, M.; Okuno, Y.; Hattori, M.; Goto, S.; Kanehisa, M. Computational assignment of the EC numbers for genomic-scale analysis of enzymatic reactions. *J. Am. Chem. Soc*. **2004**, 126(50), 16487–16498.
54. Bender, A.; Mussa, H. Y.; Glen, R. C.; Reiling, S. Similarity searching of chemical databases using atom environment descriptors (MOLPRINT 2D): evaluation of performance. *J. Chem. Inf. Comput. Sci*. **2004**, 44(5), 1708–1718.
55. White, R. H. The difficult road from sequence to function. *J. Bacteriol*. **2006**, 188(10), 3431–3432.
56. Gärtner, T.; Flach, P.; Wrobel, S. On graph kernels: hardness results and efficient alternatives. In *Learning Theory and Kernel Machines*. 2003, p. 129.
57. Kashima, H.; Tsuda, K.; Inokuchi, A. In *Marginalized Kernels Between Labeled Graphs*, Proceedings of the Twentieth International Conference on Machine Learning, Washington DC, Aug. 21–24, 2003; Fawcett, T.; Mishra, N., Eds. AAAI Press, Washington DC, 2003, pp. 321–328.

58. Mahe, P.; Ralaivola, L.; Stoven, V.; Vert, J. P. The pharmacophore kernel for virtual screening with support vector machines. *J. Chem. Inf. Model.* **2006**, 46(5), 2003–2014.
59. Swamidass, S. J.; Chen, J.; Bruand, J.; Phung, P.; Ralaivola, L.; Baldi, P. Kernels for small molecules and the prediction of mutagenicity, toxicity and anti-cancer activity. *Bioinformatics* **2005**, 21(Suppl. 1), i359–i368.
60. Ben-Hur, A.; Noble, W. S. Kernel methods for predicting protein–protein interactions. *Bioinformatics* **2005**, 21(Suppl. 1), i38–i46.
61. Helma, C.; King, R. D.; Kramer, S.; Srinivasan, A. The predictive toxicology challenge 2000–2001. *Bioinformatics* **2001**, 17(1), 107–108.
62. Kramer, S.; De Raedt, L. In *Feature Construction with Version Spaces for Biochemical Applications*, Eighteenth International Conference on Machine Learning, San Francisco, 2001. Morgan Kaufmann, San Francisco, 2001, pp. 258–265.
63. Webb, E. C. *Enzyme Nomenclature 1992: Recommendations of the Nomenclature Committee of the International Union of Biochemistry and Molecular Biology on the Nomenclature and Classification of Enzymes*. Academic Press, San Diego, CA, 1992.
64. Churchwell, C. J.; Rintoul, M. D.; Martin, S.; Visco, D. P., Jr.; Kotu, A.; Larson, R. S.; Sillerud, L. O.; Brown, D. C.; Faulon, J. L. The signature molecular descriptor: 3. Inverse-quantitative structure–activity relationship of ICAM-1 inhibitory peptides. *J. Mol. Graph. Model.* **2004**, 22(4), 263–273.

CHAPTER 7

WORKFLOW TOOLS FOR MANAGING BIOLOGICAL AND CHEMICAL DATA

THORSTEN MEINL, BERND WISWEDEL, and MICHAEL R. BERTHOLD
Universität Konstanz, Konstanz, Germany

1 INTRODUCTION

The generation of large amounts of data continues to outpace the ability of automated tools to integrate, transform, and analyze the resulting data repositories. Even though increasingly sophisticated tools are created to handle the enormous piles of heterogeneous information fragments, it is still a big challenge to support a person in extracting the knowledge nugget that in which he or she is actually interested.

In the past, complex analyzes were often performed by highly trained people who wrote and continuously updated and maintained sets of analysis scripts, written in their language of choice, such as Python or R. Such libraries were often not usable for anyone other than the original author, and continued maintenance quickly became impossible without involving the author. From a corporate point of view this was not only highly risky but also made it almost impossible to share knowledge among individuals, let alone different groups. Tools with graphical user interfaces were much more limiting in terms of analytical power but at least also enabled less well-trained users to run standard analyzes on their data. However, every time that a new data set was generated (or an existing one updated) the entire analysis needed to be reexecuted step by step. This process again made it difficult to share expertise among groups and proved to be a huge waste of time, since it required reexecution of standard analyzes over and over again. The best of both worlds is the ability to define an analysis

Computational Approaches in Cheminformatics and Bioinformatics, First Edition.
Edited by Rajarshi Guha and Andreas Bender.

process quickly and intuitively and then rerun it automatically on new data while being flexible enongh to make slight modifications, resulting in tools to truly explore and analyze data. In this chapter we discuss some of the most modern, flexible, and powerful tools to support this analytical type of data exploration: namely, tools that allow the user intuitively to model the flow of information through various processing modules. Such *workflow* or *data-pipelining* tools have emerged in the past decade or so as the method of choice for quick, flexible integration and analysis of diverse, distributed information repositories while still being able later to reuse the workflow generated to analyze new or updated data. An interesting side effect is the ability to document what has been done and to communicate this visually to other users who may not share the preference for a specific script or analysis language. As a result, workflow tools have emerged as a knowledge-capturing tool as well, allowing companies to convert individuals' experiences into valuable reference knowledge for later use.

Here we focus on four prominent representatives of such tools in the areas of chemo- and bioinformatics. In other areas, such as business intelligence and financial applications, a variety of other workflow or pipelining tools exist; however, they are outside the scope of this chapter. We discuss two commercial representatives (KDE from InforSense[16] and Pipeline Pilot[2] from Accelrys) as well as two open-source workflow tools: Taverna[22,24] and KNIME.[4,14] These four tools cover the breadth of functionality and address rather different needs: Two are targeted more toward data pipelining, whereas the other two focus primarily on the workflow organization of (distributed) tools. We go into more detail with one of the tools, the open-source Konstanz Information Miner (KNIME), as an example to illustrate and discuss the pros and cons of some of the underlying technical choices. Finally, we demonstrate how such tools can be used based on some examples.

2 EXISTING WORKFLOW ENVIRONMENTS

As we mentioned in Section it is highly preferable to model analysis tasks, as they often appear in chemo- and bioinformatics in graphical workflows. Besides the four programs already mentioned, which we discuss in more detail below, there are, of course, other tools, such as Kepler,[3,18] Insightful Miner,[17] GridMiner,[8] DataMining Grid,[12] LONI Pipeline,[19] and Triana.[34] However, they are either not commonly used, do not offer special bio- or chemoinformatics functionality, or require programming skills. Therefore, we concentrate on KNIME, Taverna, Pipeline Pilot, and InforSense KDE.

These programs can be categorized in several ways. First and most obvious is the licensing and/or development model. KNIME and Taverna are open-source projects and may be used completely free of charge, whereas Pipeline Pilot and InforSense KDE are proprietary commercial products. Both approaches have advantages and disadvantages; for new users the uncomplicated and free access to KNIME and Taverna might make them the first choice, whereas for industrial users the supposedly better and longer-term product support of the latter two may be more important.

Another fundamental property is the way in which workflows are modeled and built. A simple linear chain of nodes may already be sufficient for many tasks, but there are many others, where a node requires two different inputs and/or creates more than one output (e.g., a substructure filter which obtains the substructures at the first input and the molecules to be filtered at the second input, subsequently placing the matching molecules at the first output and the nonmatching at the second). Therefore, the theoretical model underlying the workflows is usually a directed acyclic graph, meaning that nodes can have several inputs and outputs, but loops are not allowed. This does not, however, mean that the concept of loops (i.e., iteratively executing parts of a workflow with different data or parameters) is not possible. It is just the data that flow strictly from one end to the other.

Once the workflow has been designed, data are fed into it. How the input data are processed is another conceptual design issue. KNIME, for example, is table based (i.e., the input data are processed completely by any node in the workflow before the results are made available to successor nodes. This may seem to be a disadvantage over the other concept, where single data rows are sent along the pipeline as soon as they become available—Pipeline Pilot uses this concept. However, quite a few processing elements in the workflow still require all the input data anyway (e.g., all kinds of machine learning/data-mining algorithms, such as clustering, decision tree builders, or visualization nodes). Therefore, the implicit parallelism achieved by row-oriented pipelines is only of limited use. In the case of Pipeline Pilot the drawback of this concept is that the processed data are not stored at each node's output, only at special nodes that collect the results. In KNIME the result table of each node is available at all times, even after closing and reopening a workflow (provided that it has been saved, of course). Another disadvantage of pipelining is the limitation to just one input port to nodes; therefore, filtering an input table (or data stream) based on references from another source (e.g., filtering a set of molecules based on a set of substructures) is not done easily. Taverna and KDE operate similar to KNIME in that a node must first finish with all data before its successors are executed. Apart from these conceptual differences between the four programs, there are, of course, many other differences, which we discuss further in the following sections.

KNIME

KNIME is a modular data exploration platform that enables the user visually to create data flows, selectively execute some or all analysis steps, and later investigate the results through interactive views on data and models. KNIME was developed (and continues to be expanded) by the Chair for Bioinformatics and Information Mining at the Universität Konstanz in Germany. KNIME is based on the Eclipse platform, and thanks to its modular API, is easily extensible. Customs nodes and types can be integrated quickly, enabling KNIME to be used not only in production environments but also for teaching and research prototyping. Apart from a large collection of nodes for data preprocessing tasks, it also contains many nodes for data mining and visualization. As already mentioned, KNIME workflows are built from nodes that can be connected in a DAG (directed acyclic graph)-based structure (i.e., a graph without loops). Nevertheless, looping support is available through special loop start and end nodes, which execute the nodes between them (the loop body) a certain number of times or until a specific criterion is fulfilled. This can easily be used for common tasks such as cross-validation, feature selection, or parameter optimization (the loop concept is explained in more detail in Section 3.3). Such subworkflows can be extracted into *metanodes*, which may then be reused in other workflows.

Another useful concept for interactive data analysis is the *hiliting* feature, which marks related data rows in all branches of the workflow. We can, for example, consider a node that finds frequent fragments in a database of molecules, such as KNIME's MoSS[6,7] node. Once an interesting fragment has been found, our interest focuses on the molecules that contain this fragment. By highlighting one (or more) rows in the interactive table, which is connected to the MoSS node's fragment output port, the corresponding rows in the second interactive table, which is connected to any node before MoSS, will also be highlighted. This concept is known as *visual brushing* in the data-mining world. Figure 1 shows part of this type of workflow, together with the two table views and some hilit fragments or molecules.

Connections in KNIME workflows do not only transport data tables, but may, in fact, transfer arbitrary "port objects" from one node to another. This might include trained predictive models, such as decision trees, support vector machines or regression coefficients, and database connections

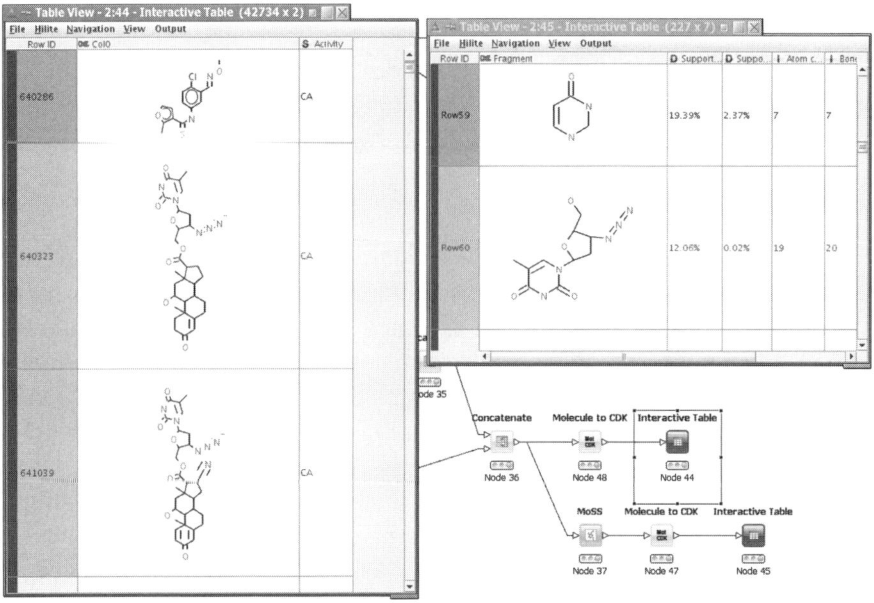

FIGURE 1 Parts of a chemoinformatics-related workflow, demonstrating KNIME's hiliting feature. The MoSS node searches for frequent fragments in a set of molecules. Selecting and highlighting several of the fragments found after the MoSS node will also highlight the corresponding rows in the workflow's upper branch.

(again, more technical details are provided in Section 3). The various connection types are indicated by different colors and shapes of the connectors attached to the nodes. KNIME has been designed as a generic data analysis and visual exploration platform and therefore does not offer special bio- or chemoinformatics functionality in its base distribution, which somewhat comprises nodes for data input and output, row- or column-centered manipulation of data tables (filtering, sorting, shuffling, partitioning, etc.), and data visualization (scatter plotter, box plot, line plot, etc.) as well as data mining. However, as KNIME is built on the Eclipse[13] framework, it is simple to add new functionality—usually in the form of new nodes—by providing additional plug-ins. The official KNIME website offers several extensions, the most important (in the scope of this chapter) being several chemistry-related data types (e.g., SDF, Mol2, SMILES) and corresponding input/output nodes and the CDK (Chemistry Development Kit[9,30]) plug-in, built on it. By integrating parts of CDK's functionality, molecules can be visualized as two-dimensional (2D) structure diagrams or—if available—as three-dimensional (3D) structures in table views, various properties and descriptors can be calculated and substructure searches can be performed. Figure 2 shows a workflow with

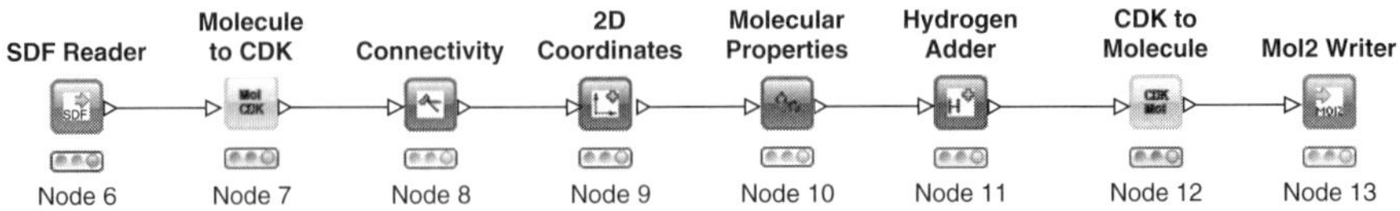

FIGURE 2 workflow showing some of the CDK nodes available.

some available CDK nodes. Other integrated tools are OpenBabel and the already mentioned fragment miner MoSS. In addition to these open-source nodes, several commercial extensions are available. Schrödinger[27] provides a number of nodes offering access to their chemo- and bioinformatics applications such as LigPrep, Glide, or BLAST from KNIME. Tripos[33] offers nodes for, among others, the generation of chemical fingerprints, 2D alignments, and conformations, as well as nodes that integrate different DB cartridges. Infocom[15] has integrated several of ChemAxon's tools,[11] such as database access, several property calculations, and R-group decomposition (for academic use, these nodes are currently free of charge). Symyx[31] has developed several free nodes for reaction-based enumeration, stereonumeration, or R-group decomposition (however, some of them require a Cheshire license). Molecular interaction fields and chemometrics in pharmaceutical discovery are provided by several nodes from Molecular Discovery.[20] Treweren Consultants[32] has packaged its THINK nodes for de-novo derivative generation, structure-based virtual screening (docking), and searching for novel actives using pharmacophores with volume constraints. Several other software providers are working on the integration of their respective tools.

Besides this huge collection of special bio- and chemoinformatics nodes, there are other extensions that may be of use in the workflows, such as basic reporting functionality via the BIRT project,[5] the integration of the R statistics package,[25] and Weka[35] for a vast collection of machine learning algorithms. One of the next releases will also feature several image- and text-processing nodes.

Taverna

Taverna is another free, open-source software tool that helps users to create scientific workflows, primarily in bioinformatics. Taverna is being developed as part of the UK's myGrid initiative.[21] Its main focus is on using independent Web services via a common workflow-oriented front end. Nodes (also known as "processors" in Taverna) are discovered dynamically upon startup by querying pre- and user-defined Web service registries. Together with several local nodes, such as input/output, string operations, or dynamic workflow controls, they are shown in a structured node repository. The user can subsequently build a workflow (which is a directed acyclic graph, as in KNIME) by first adding nodes to it and then connecting their various input and output ports. In contrast to KNIME, this cannot yet be performed directly inside the graphical workflow layout but must be done via the list of processors, as the graphical layout is simply a picture generated by GraphViz (this is supposed to change in the next major release). The workflow is represented internally in the Scufl workflow language and can in principle be run separately from Taverna. The use of Web services as main data-processing components has the advantage that the user does not have to install additional software components; instead, they are used out-of-the-box. As a side effect, the workflow can use the remote computing infrastructure, which is usually more powerful than the local computer. Databases and other resources can also be used. This does not come without a cost, however. Changes at the remote side can easily break existing workflows. Also data must be transferred to the remote side first. The result is then transmitted back, which can become a major problem if the data sets are large. Also, Web services may be temporarily unavailable or may simply be removed, which then renders the workflow useless. The data passed along the connections in the workflow do not follow a strict schema like the data tables in KNIME, which are strongly typed (each column has a distinct type). Instead, Taverna uses mime types as "data blocks." More advanced data structures can be built by defining lists or trees of mime types. Unfortunately, this generality makes it impossible to check if the input/output data of two connected nodes matches before executing the workflow. For converting data in such cases, Taverna offers numerous "shim" services, which make sure that two nodes can be used together.

Besides data connections, Taverna workflows usually contain coordination links between nodes. These are useful if a node needs the results from another node but where the results are not transferred by data connections but are stored directly on the remote side (e.g., inside a database). A coordination link between these two nodes ensures that the second does not start before the first is finished. Running a workflow in Taverna always entails executing all nodes from beginning to end, whereas other programs often support execution of parts of the overall workflow as well.

Pipeline Pilot

SciTegic Pipeline Pilot[2] is a commercial pipelining tool and—given its wide use—probably the market leader in the chemoinformatics area.

Components[†] for data retrieval, filtering, basic analysis, and reporting are available. Although it is known primarily for its chemoinformatics capabilities, it also provides add-ons for, for example, bioinformatics (gene expression profiling, sequence analysis), image analysis (as used, e.g., in high-content screening), chemically aware text mining, and statistics and data modeling. Pipeline Pilot realizes the *pipelining paradigm*; that is, individual data entities (which correspond to data rows in KNIME, e.g., compounds) are passed on to subsequent components as soon as their processing is finished in one component. There are two principal advantages of this pattern compared to the workflow paradigm, which is implemented by KNIME or Taverna, for example: It implies high scalability by design since no intermediate results need to be cached, and second it allows for an early inspection of the final results. Both benefits come in handy if the entire analysis flow can be streamed (i.e., the components do not iterate the data multiple times, and the focus of analysis is on the final results only). However, this pattern is less well suited for tasks that require multiple iterations on the data as required by many preprocessing or data-mining algorithms. It has also shown to be less convenient if subsequent modifications to the pipeline are necessary; for example, the user decides to subbranch from the pipeline or to modify any component in between; this typically requires a rerun of the entire protocol.[‡]

Pipeline Pilot uses a client–server architecture in which there are three different client options, which differ in available functionality and commercial support schemes: a professional client for "own" protocol creation and customization as well as protocol sharing, a "lite" client for "own" protocol creation, and a Web port for convenient deployment of existing protocols.

Similar to KNIME, Pipeline Pilot enables a seamless integration of third-party tools. This can be as easy as calling an external executable with dedicated command line flags, the integration of different databases through ODBC protocols, or the call of external Web services. It also allows for a fully fledged integration of libraries using customized components. The latter is commonly used by different independent software vendors, such as ChemAxon, Tripos, and Molecular Networks (the Web page currently lists more than 30 partners). Pipeline Pilot is a commercial software product that addresses many user needs, such as client-server

[†]The term *component* translates to *node* in KNIME or *processor* in Taverna nomenclature, whereas a workflow is known as a *protocol* in Pipeline Pilot.

[‡]There are components that allow a buffering of intermediate results in a "finish here, resume later" manner, although these components must be placed in the pipeline beforehand.

architecture, role management, Web service integration, and reporting; however, it has drawbacks with respect to usability and is also a relatively expensive solution.

InforSense

InforSense's workflow editor is part of a larger suite of applications that targets a wide range of applications: from business intelligence to customer relationship management, and from fraud detection to the life sciences. InforSense uses the workflow engine as the central backbone to integrate various different modules, forming the *InforSense platform*. Some of the resulting applications wrap the workflow so that the user can concentrate on the resulting views: for example, the reporting analytics product.

From the standpoint of this chapter, we are interested primarily in the workflow authoring and execution tool. Similar to most of the other tools discussed here, the editor allows a pipeline of modules that read, manipulate, model, analyze, and finally, write data to be assembled visually. One strength of this platform is the ability to tightly integrate other tools, which is demonstrated by the large list of partners who integrate access to (some of) their tools into this workflow engine, especially in the chemoinformatics domain. However, access to databases and calls to in-database data analysis routines are also supported.

InforSense uses an object-based design paradigm, which can encapsulate data tables as well as more complex objects. This paradigm results in data processing similar to KNIME (i.e., nodes tend to be executed fully before successors become executable). Again, this does enable easy integration of more sophisticated nodes, which rely on seeing the data in its entirety (or repeatedly) in order to, for example, build a model of the data, such as a decision tree or a rule model to predict biological activity.

InforSense's product portfolio encompasses a variety of different extensions, based on this workflow engine. ChemSense, to name one, adds chemoinformatics functionality and allows access to chemical compound databases, specialized data cartridges, and modeling tools from several vendors. Built-in functionality offers basic property calculations and substructure analysis in addition to the visualizations useful for SAR analyzes. Extensions allow text and images and also biological data such as gene and protein expression profiles to be analyzed. However, it is clear that InforSense's expertise lies in the analysis and visualization of complex scientific data sets, while with regard to life science applications, the integration of appropriate partner tools is essential. This also becomes apparent when browsing through the list of industries using this platform. In addition to pharmaceutics and life science, diverse areas such as financial services, manufacturing, and communications illustrate the breadth of areas covered.

InforSense offers a commercial product with a wide range of applications. Using partner extensions, a full suite of chemo- and bioinformatics solutions can be addressed; however, to be useful as-is tools, the standalone tools from InforSense lack domain expertise.

3 TECHNICAL ISSUES

In the following we describe more technical considerations to be made when designing workflow or pipelining tools, using KNIME as an example. Most of these considerations were inspired by how users expected workflows to behave, whereas others resulted from design decisions that required development in KNIME to be as easy as possible while still enforcing the stability of third-party contributions.

The Node Concept

As already pointed out, the most important part of each workflow are the data-processing elements, the nodes (Figure 3). Usually, a node has one or several input ports and one or several output ports. Some nodes, such as file readers or writers, have only input/output ports. A node in KNIME consists of up to five components:

- The NodeModel, which processes the data.

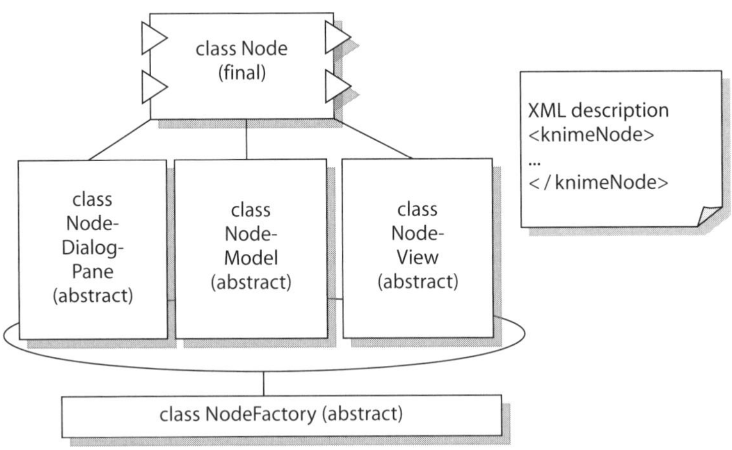

FIGURE 3 A node is made up of the five components shown. NodeView is an optional class.

- The (optional) NodeDialog, which lets the user enter the settings for the node's operation (e.g., which column to use, parameters for the algorithm).
- The NodeFactory, which creates instances of the model, dialog, and view (see below).
- An XML file, which describes the nodes functionality, the parameters the user can adjust, what type of input it expects, and what type of output it creates.
- Optionally, one or more NodeViews, which visualize data or the model built inside the node (e.g., a decision tree or a lift chart).

These are the abstract classes that a node implementor has to extend. Everything else, such as data transfer, invocation of the model, showing the dialog, and so on, is handled by the framework. The most important class, called NodeModel, has two main functions:

- To check if input table structures and user settings are compatible for the specific operation and, if so, to determine the structure of the output table and report it to the successor nodes. This step is performed before any actual data are processed and is called the node's *configuration*.
- To take the input data, run computation, and create the result table. This step is called its *execution*.

Separating configuration and execution makes it possible to build a workflow, configure all its nodes, and then afterward, begin (the potentially long-running) processing the data while being sure that output and input data of all connected nodes will match (there are some rare cases in which a node cannot determine the structure of its output table before having seen the actual data). The NodeDialog consists of one or more Swing panels, which can contain other possible Swing elements. Usually, the user adjusts a few settings based on the structure of the input table, which are in turn transferred from the dialog to the model upon configuration or execution. The optional NodeView is also a user-defined Swing component which has access to the NodeModel and can visualize its internal model or the input data (e.g., Scatter Plotter). While executing a workflow the user is usually interested in its current state. Therefore, the status of each node is displayed (waiting for execution, executing, executed, unable to execute). If it is running currently, progress is indicated by a progress bar. Of course, some nodes cannot make such an estimation, but they can at least report how many rows have been processed, or how many solutions (output rows) have already been found. In any case, the user is able to cancel the node's execution at any time.

Data Handling

Three major design goals of KNIME have been scalability, ease of use, and a simple extension mechanism. All three principles raise specific challenges to the data-handling mechanism used in KNIME. These challenges and their solutions are discussed in the following sections. To understand them, we first take a brief look at the data structure that is used to pass data through the workflow.

The standard data format is table-based, whereby rows represent different records and columns describe record attributes. As an example, consider a data set of chemical compounds (e.g., represented by SMILES i.e., strings), which are described by numerical measures such as log P, molecular weight, and atom count. The SMILES strings and each of these attributes are represented by individual columns, whereas a row represents a specific compound with its specific attributes. Figure 4 depicts the general architecture and the access methods of this data structure (i.e., the *BufferedDataTable*). Each table has a predefined table specification (*DataTableSpec*), which consists of typed and labeled columns (*DataColumnSpec*). The data, which comply to this specification, are accessible via an iterator (there is no random access on records), which returns records one by one (*DataRow* elements). The fact that the data are only exposed using an iterator already highlights the fact that tables may have an almost arbitrarily high number of rows (random access in a large collection would have been prohibitively expensive), whereas the column count is typically a small number (ranging from very few to a couple of thousand). KNIME uses a smart caching strategy to swap tables that are too large to fit into main memory. Subsequent iterations on these tables read from the hard disk so that the table is never kept in main memory as a whole (unless a specific implementation of a node decides to do so). KNIME follows a philosophy according to which (temporary) disk space is not a limitation, whereas main memory can be.

Efficient Data Referencing KNIME realizes a workflow pattern that buffers each node's computed outcome. This allows for an inspection of intermediate results (the user can review what has been computed at each individual node even after the entire workflow has finished executing), and second, it offers the possibility of continuing with newly created subbranches after successful execution without the need to rerun the entire workflow. This buffering strategy can be quite expensive in terms of memory or disk usage since many nodes duplicate their input if not handled with caution. KNIME uses table references for nodes that perform column-based transformations to input tables; that is, it does not copy the input data

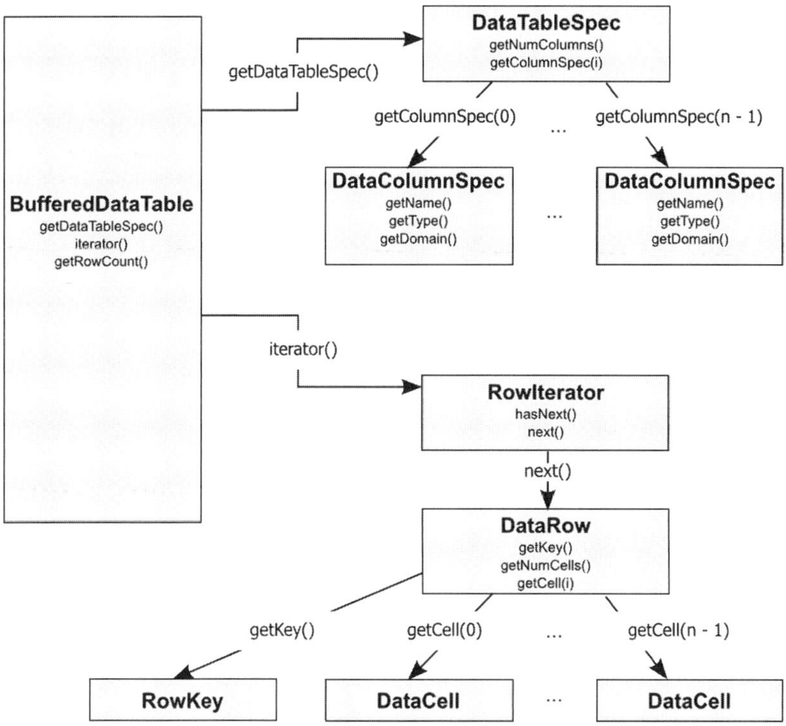

FIGURE 4 KNIME uses BufferedDataTables as the underlying data structure. It has a fixed number of typed columns and a (possibly very large) number of DataRows. Each row represents a record in the data set.

but references them and saves only what has changed. In practice, it turns out that a vast majority of nodes that actually perform these types of modifications (i.e., they do not modify the entire input data but instead append, replace, or modify individual columns). Examples are preprocessing nodes (e.g., column filter), nodes for (chemical) property calculation, and type converters that parse, for instance, numbers or translate chemical representation from one format into another. It is important to note that successor nodes iterating the output of such column transformers will not notice the specific instance of a BufferedDataTable; instead, they iterate their input, whereby the underlying iterator assembles the "right" data on the fly.

There are other modifiers to which a similar trick can be applied. However, we do not want to delve into too many technical details here but restrict ourselves to brief mention of the following:

1. *DataTableSpec replacement*. Some nodes perform no data manipulation at all but only change the table specification (e.g., nodes that

rename columns or set certain properties for them). These nodes do not create any data but only wrap the input table while changing its specification.

2. *Concatenation*. Nodes that process the input data in chunks in order to exploit today's multicore architecture (see the "Parallel Processing of Data Inside a Single Node") Section make use of table concatenations. That is, individual workers create only parts of the final output table, which are eventually combined to create the output table.
3. *Column Append*. Specifically, the nodes mentioned in the preceding item often only create new columns and hence make use of the concatenation scheme. To attach the final result table to the input table, they implement an append operation. An illustrative example is a node that generates 3D coordinates for structures (a computationally rather expensive step which can easily be parallelized); the result consists of the input table joined with the concatenated result of the individual workers.

All these operations can be nested arbitrarily, which provides a very powerful framework to reduce the number of data stored in each node. Unfortunately, this trick cannot be applied to nodes that perform row-based transformation (e.g., sorting or shuffling the data rows), as this changes the order of the rows. For these special cases, the data actually need to be copied. Luckily, there are only a handful of scenarios where this becomes necessary, as most nodes perform calculations or modification based on individual rows.

New Data Types Another design principle of KNIME is its extendability. In Section we discussed the definition of new nodes that perform customized operations. However, this is not the only direction further extension can take. The definition of new data types is often of interest if the focus is on analyzing objects of a very specific nature. Very prominent examples are molecular structures, proteins, and reactions. As a matter of fact, KNIME in its base version has no chemistry support; standard types are limited to a very small set that includes strings, real and integer numbers, bit vectors, and cell collections. Chemistry functionality is available in additional add-ons (called plug-ins), which provide chemistry data types (such as definitions for SMILES, Mol2, SDF, or PDB[†]) and/or nodes that consume or produce data of these nonstandard types. These nodes would be useless without a definition of the respective data type.

[†]In comparison to Pipeline Pilot, KNIME does not define an "own" chemistry type.

There are more examples of customized data types, which we also discuss in Section 4, including image data, large text documents, and 3D object representations.

The definition of a new data type is simple. It boils down to the definition of common access methods via an interface (extending an interface DataValue), enabling node implementations to retrieve the cell internals, and concrete implementation of a container (called DataCell). The DataValue interface additionally defines common properties of the type: for instance, a comparator, which is used when a table is to be sorted according to a column containing cells of the new type. It can further define a renderer class that is, the functionality to visually depict an object in KNIME views (e.g., to show a 2D depiction of a molecule within a table viewer). DataCell implementation, on the other hand, is used to represent the content and (possibly) to define a custom serializer, which is used for efficient storage and restoration of objects of that type (otherwise, the slower standard Java serialization is used).

Some cell objects, such as proteins, images, or documents, can be relatively large (up to a couple of MBs) and need special handling to ensure fast data iterations and efficient disk usage. These types can be represented by *blobs* (binary large objects); that is, they are not treated like the usual cell objects, which are saved consecutively in a data stream, but instead are saved separately on the hard disk. Further reference to this cell object does not actually store the cell content but, instead, only address information, to ensure that the content is not duplicated. This also has the advantage of enabling faster iterations on tables containing blobs, as they do not get restored as part of the interpretation of the data stream but only on when the cell is actually accessed. The blob concept has shown to be very useful in practice, specifically in the domain of image and text mining (see Section 4).

Loop Support

We have already mentioned in Section 2 that the workflow structure in all tools discussed here is usually a directed acyclic graph (i.e., there are no loops). Still, some type of iteration concept can be of use in many applications: from common tasks such as cross-validation or feature elimination, to iterating over a list of input files, to even more sophisticated approaches such as using the rows from an input table as parameters for one iteration of the loop (e.g., to find the optimal sets of parameters). Before KNIME 2.0, this was only possible by using the batch mode, where node parameters are set on the command line before KNIME begins. A "loop" is then simulated by executing the same flow repeatedly with different starting parameters. This approach, however, has several drawbacks. First, for each "iteration," KNIME is restarted from scratch, which takes some time. Second, setting the right parameter on the right node via the command line switches is not very intuitive because one the nodes' XML configuration files have to be examined, and third, more complex loops such as feature elimination are not possible at all, because each iteration depends on the results from previous iterations. Therefore, since version 2.0, loops have been supported directly by the workflow manager.

A loop consists of two special nodes, the loop start and end nodes. They are special in the sense that they have access to each other (usually, a NodeModel cannot access any other NodeModel) and that in contrast to the nodes inside the loop, they are not reset while the loop is executing. This is important, because they have to maintain their state during the iterations. For example, the start node of a cross-validation loop has to keep track of the data table partitions that are used as training and test sets, while the end node has to collect the results from each iteration. The decision as to whether the loop should be stopped or another iteration performed is made by the end node. This means that the loop is executed at least once ("do-while" loop). Figure 5 shows the parts of a workflow that contain a feature elimination loop. Conceptually, feature elimination consists of two nested loops (the user sees only one loop in the flow). The "outer" loop is executed as many times as there are features. After each iteration, one of the features (the one that was least important for the trained model) is left out. The "inner" loop determines which of the remaining features is to be removed next by the outer loop. The end node collects the model performance for each "inner" iteration and then decides which attribute is left out for the next "outer" iteration. This process is repeated until only one feature is left. The end node then builds a special feature elimination model, which can be used later by a feature elimination filter, which enables certain columns to be filtered from its input table based on the performance of the model inside the feature elimination loop.

The loop concept becomes even more powerful when it is combined with flow variables. These are simple key-value pairs, which are passed along the connections. Each node may read or write these variables and use them during execution. Inside the dialog the user can then assign a certain variable to a dialog setting. When the node is executed, it takes the value of the variable instead of the value entered in the corresponding dialog component. The last missing piece is a loop start node which takes a data table as input, holds different parameter settings for the node(s) inside the loop, and takes one combination (= row) in each iteration.

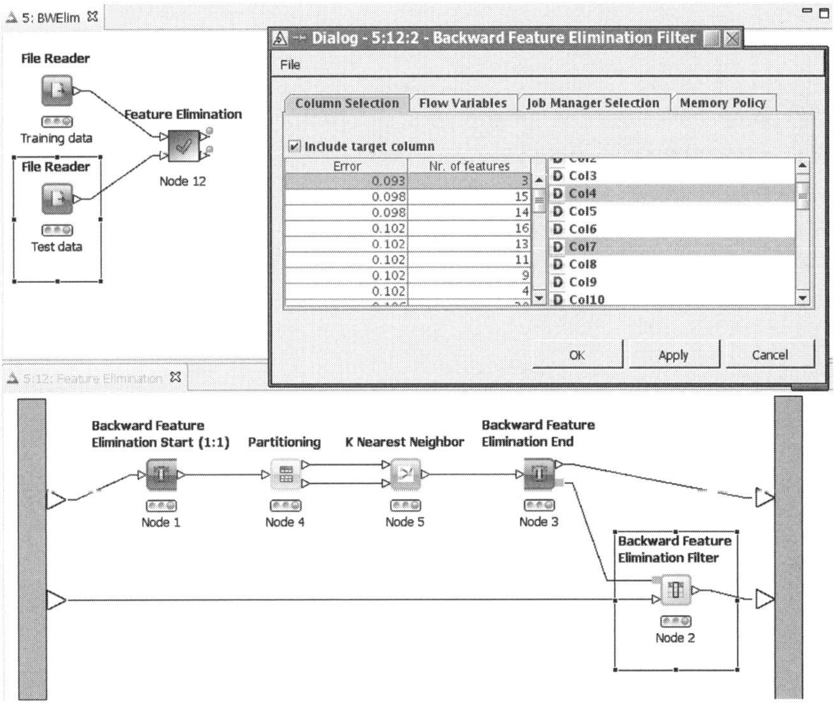

FIGURE 5 Workflow containing a feature elimination loop. After the loop has been fully executed, the feature elimination filter can be used to filter out columns from another table, based on the error rates seen while looping.

Parallel Data Processing

In the field of workflow systems, there are several ways to exploit the parallelism offered by multicore processors or multiprocessor systems with shared memory. The simplest and most obvious is the parallel execution of different branches in the workflow. Each node that is ready for execution can run in its own thread (and is executed on any free processor or core). In the case of real data pipelines (as in Pipeline Pilot), even nodes depending on each other can run simultaneously because as soon as a node has produced one result row it is sent to its successor nodes, which immediately process it further. However, as we explained in Section 2, this inherent parallelism cannot be used throughout the entire workflow. Also, if the runtime per row is very different for the nodes involved, the achievable speedup is quite limited.

In KNIME, the next, more advanced approach is to process the data rows of the input table in parallel. Finally, the most sophisticated method of parallelization is to execute entire subworkflows in parallel (e.g., the different iterations of a cross-validation). In addition, some algorithms allow for parallelization themselves; however, this aspect is usually quite independent of the workflow. Next we explain the various approaches in more detail and also take a brief look at what a programmer must do to write a node that is capable of processing data in parallel.

Parallel Execution of Independent Nodes As already mentioned, the most obvious method of parallel data processing is to execute several independent nodes at the same time. To avoid overloading if too many nodes are ready for execution, KNIME uses a thread pool that is limited in size and reuses threads (Figure 6). The user can specify how many threads should be used at the same time. From the programmer's point of view, nothing needs to be done to allow parallel node executions. They are handled automatically by KNIME's workflow manager; it keeps track of queuing and executing the nodes in a workflow.

Parallel Processing of Data Inside a Single Node A considerable number of nodes (especially preprocessing nodes) perform computations based on single rows independent of other rows. Shown in Figure 6 are examples of this type of node which parse molecular representations and convert them into internal representations (e.g., the SDF parser) or nodes that manipulate image data (e.g., low-pass filter or Voronoi segmentation). In KNIME these nodes are called *threaded nodes*, and implementing them is not very different from implementing normal nodes. The framework takes care of splitting the input table into chunks; each chunk is processed in a separate thread, which is taken from the thread pool mentioned earlier.

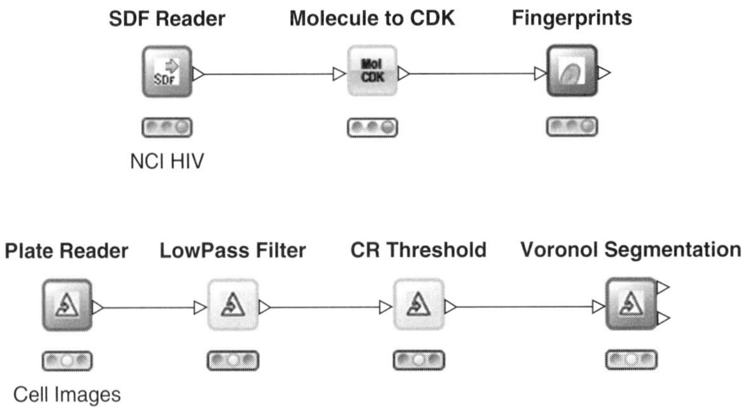

FIGURE 6 Two small workflows with threaded nodes.

To achieve an equally distributed load among all threads on the one hand and a low overhead on the other, a suitable balance between the size and number of chunks is important. Currently, four times as many chunks are created as available threads in the pool. The abstract model provided by the framework calls a method in the concrete subclass for each row of the input table, which then returns the new cells that are appended to the row. In the final phase the results are merged (cf. the Section "Data Handling") and a complete output table is built. The programmer need only be aware that the code is called by several threads concurrently. Therefore, synchronized blocks should be avoided and write-access to common data must be used carefully. Apart from that, implementation resembles the normal node's API. Almost the same usage model applies in cases where the number of output rows differs from the number of input rows or when the structure of the output table is completely different from the input table structure. Again, the framework invokes a special implementation for each row of the input data. This time, however, a (possibly empty) set of complete rows must be returned. Again, the framework takes care of merging the final results. The threaded nodes have a slight overhead that comes from splitting the input tables and merging the results. This depends largely on the size of a row (the number of columns and the size of the objects in the cells) and on the input/output speed. Generally, however, this overhead does not significantly impair performance.

Besides this local thread-based parallelization, we are currently working on distributing the different chunks as jobs into a local cluster. The user can select how many subjobs should be created (or the number of rows per job), and the framework keeps track of splitting the input table, creating the cluster jobs, and finally, merging the results. Behind the scenes a small workflow is created that contains just the node(s) that should run on the cluster. On the cluster, KNIME is started in batch mode and runs the workflow with one of the input table chunks. Ideally, this distributed execution should offer the same "feeling" as locally executed nodes (i.e., it should be possible to cancel them and proper progress should be reported to the user). Also, users should be able to close their KNIME session and reopen it later and still obtain the results from the cluster jobs (if they have already finished) or monitor the current progress.

Parallel Processing of Subworkflows Earlier we explained the concept of loops that repeatedly execute parts of a workflow. A popular use case is cross-validation to estimate the stability of a trained model. Figure 7 shows an example of the use of a cross-validation node. First, the data are read and then fed into the cross-validation loop. Inside the loop the X-partitioner is responsible for splitting the data into training

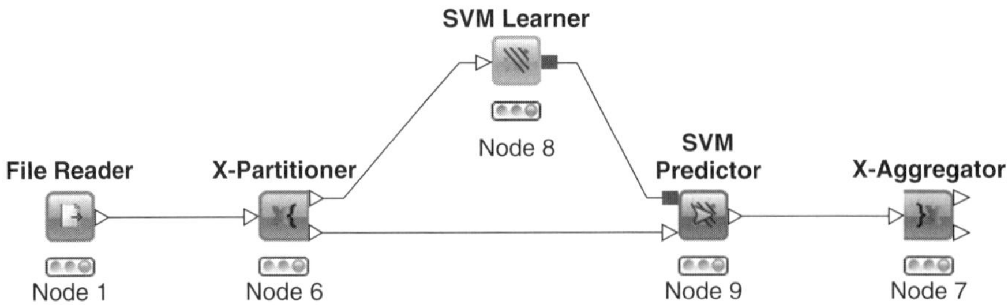

FIGURE 7 Workflow with a cross-validation node and its internal subworkflow.

and tests sets, whereas the X-aggregator node collects the results and creates statistics for each iteration. Inside the loop, the user has to insert at least two nodes that build a model based on the training data and classify the test data using the model. In the example workflow shown, the SVM learner node is used, which trains a support vector machine.

Using support vector machines is relatively time consuming, even for small data sets, and becomes increasingly so if there are 10 or more iterations in a cross-validation node. In such a case, all iterations are independent of each other, and partitioning into training and test data must be identical (i.e., in each iteration another 9 of 10 partitions are used for training, and the remaining partition is used for testing). Therefore, it is quite natural to parallelize the execution of single iterations. Unfortunately, this time, more effort is needed from both sides, the KNIME framework and the node programmer. The fact that a single node is now run in several threads at the same time, especially, poses quite a lot of problems. That these difficulties are solvable has already been demonstrated in a prototype implementation.[29] The fact that this requires some major changes to the internal execution strategy of workflows and to the node implementations themselves has kept us from integrating it into the official KNIME releases so far.

Integration of Third-Party Contributions

We mentioned in Section 2 that besides the CDK integration, the base KNIME distribution does not offer much bio- or cheminformatics-related functionality. Most of the nodes available come from third-party vendors. Because KNIME is based on Eclipse, it is very easy to add additional nodes (or other functionality) by using Eclipse's plug-in mechanism. The KNIME core offers *extension points* to which other plug-ins can be attached. The plug-in offering the extension point subsequently reads out all attached extensions; in the case of KNIME, these are usually additional nodes. The nodes are then added into the node repository automatically and are usable like any other KNIME node. This extension mechanism allows for a lot more than simply being able to add nodes. In the current development version a plug-in can even change the way that single nodes are executed by providing a special node executor. The distributed processing of a node inside a cluster that we described earlier is realized by an additional plug-in.

Quite often, users and vendors want to integrate existing programs or scripts into KNIME. For very simple scripts, KNIME offers the external tool node, which creates a CSV file from the input table, launches an external program with this file as an argument, and finally, reads a result file back in. For small amounts of table-structured data, this works well, but it does not work for large numbers of molecules in SDF format, for example (which has a special internal structure). In such cases, dedicated nodes have to be programmed to create the program's input files in the right way and to parse the results in the correct manner back into KNIME. However, this file-based transfer of data is far from optimal. Ideally, the external program could work inside a "pipe" created by the KNIME node. It is given input data on its standard input and writes the results to standard output, whereupon KNIME reads them back in. This not only renders the intermediate files obsolete but provides for much better progress reporting, as KNIME simply counts the rows it has already piped into the external program (or the number of results it has received). However, usually the external program has to be adapted for this type of piped processing. Unfortunately, currently, most vendors refrain from modifying their existing software. However, with regard to open-source software, this should be a manageable task.

4 EXAMPLE FLOWS AND USE CASES

In this section we show some example workflows that come from three different research areas. The first, a chemoinformatics-related use case is the realization of a classical virtual high-throughput screening protocol; the second is a bioinformatics application, the automated analysis of cell assay images; and the third deals with the more general issue of text mining in scientific publications. The workflows presented are all implemented with KNIME, but in principle, similar workflows should be possible in the other workflow environments as well.

Virtual High-Throughput Screening

In traditional high-throughput screening (HTS), hundreds of thousands of compounds are automatically tested for their reaction in combination with a given target protein. Usually, the degree to which a compound inhibits the activity of the protein is measured. Molecules with high inhibition rates are then tested further in smaller assays, where the exact concentration at which they prevent 50% of the protein's activity (the IC_{50} value) is determined.

HTS is quite an expensive and time-consuming process which severely restricts the number of compounds that can be tested. Therefore, quite a few computational methods have been developed that try to predict a molecule's activity without performing real tests. Common approaches are:

- Protein–ligand docking, where a compounds is "docked" into the protein's active site and the total energy of the bound ligand-protein complex is taken as a measurement for the assumed activity (the lower the energy the better it should bind to the protein).
- Several ligand-similarity methods, such as Feature Trees,[26] Cofea,[1] or Shape Signatures,[36] that compute a similarity measure based on the molecules' properties, driven by the assumption that similar molecules show similar reactions.
- Pharmacophore searches, where abstract properties (charge, volume, proton donors/acceptors) of known active (or inactive) compounds describe a 3D search pattern with with which the entire database is scanned.
- Simple target-independent molecular descriptors, which can also be used to compute the similarity between molecules.

These approaches can now either be used to replace HTS (which is usually *not* applied) or as a preselection step in which a set of compounds from a virtual library is selected that is processed afterward by traditional HTS. Figure 8 shows a workflow for the second approach.

On the left, data are read into the workflow. This not only includes the molecular structures, which are later processed by some Schrödinger nodes, but also text files containing output from a Feature Trees comparison and from Cofea. This can be slightly more complex, as in the case of the feature tree similarities. In the lower part of the workflow, some of the Schrödinger nodes essentially perform a docking of the virtual library into a prepared protein. Then the scores and similarities from the various sources are joined together so that each row corresponds to one molecule and its attributes are in the columns. The following meta node computes a number of very simple combined scores by taking the maximum (or minimum) scores of a certain approach (e.g., the maximum of all Cofea

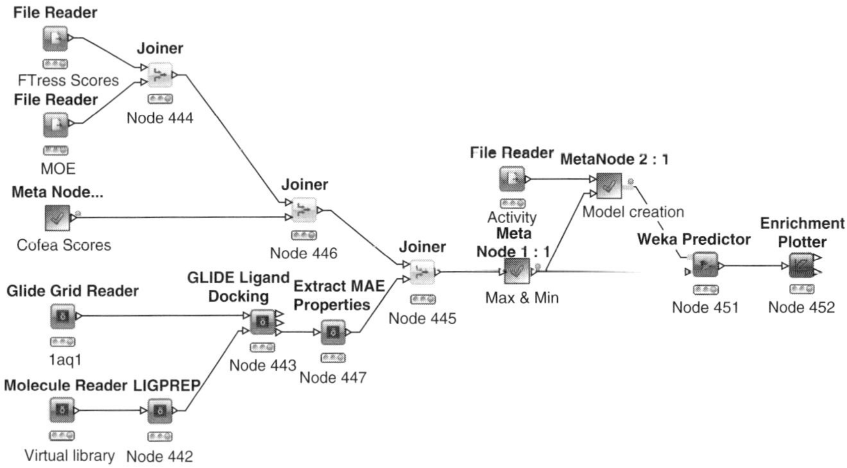

FIGURE 8 Workflow used for preselection compounds from a huge virtual library prior to HTS.

scores from comparisons to various references). Then another meta node follows, which contains a cross-validation loop (see Figure 9). This part of the workflow is used to measure the quality of the built model if activities for some compounds are already known. A support vector regression algorithm is used to predict the IC_{50} values, which later can be used to sort the molecules. The cross-validation loop splits the input data into training and test sets, uses the training set to build the model, and predicts the value for the test set. The loop's end node collects the results from

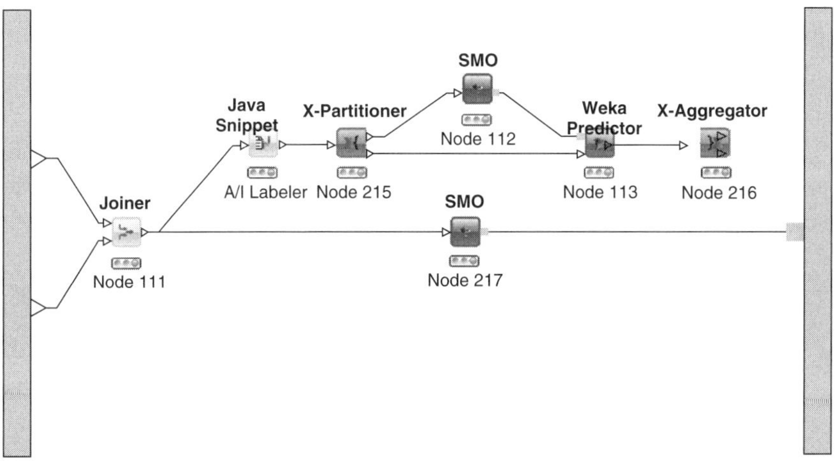

FIGURE 9 Cross-validation loop used to estimate the quality of the trained model.

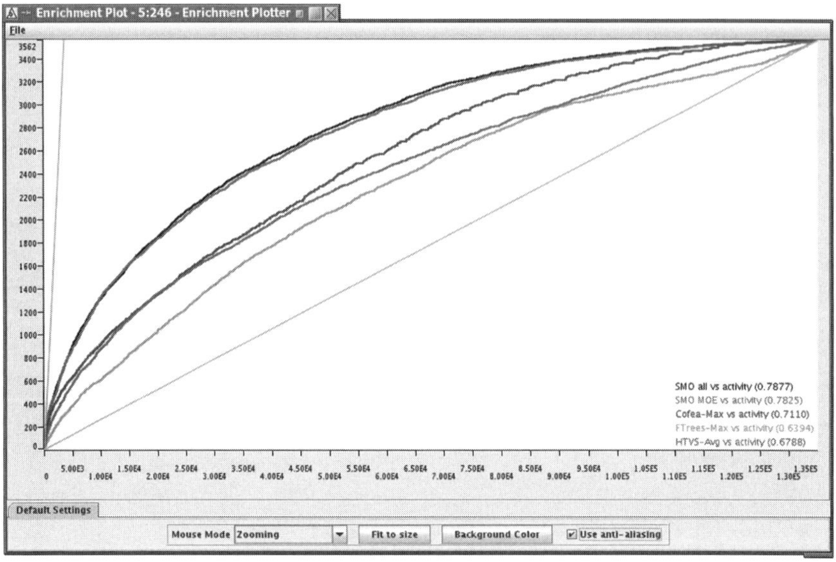

FIGURE 10 The enrichment plotter shows all molecules on the x-axis and the active molecules on the y-axis. The different graphs are created by first sorting all molecules based on the prediction and then plotting the number of actives discovered among the first x ranked molecules. The light-gray line on the left indicates optimal enrichment whereas the diagonal indicates random enrichment.

each iteration and finally creates a statistic (error rate) for each iteration. The prediction results can subsequently be visualized graphically via the enrichment plotter (see Figure 10) as well.

Cell Assay Image Analysis

The second example workflow we present is used to analyze cell images. In the last few years the development of high-throughput imaging instruments (e.g., fluorescence microscope cameras) resulted in them becoming a promising tool to study the effect of drug candidates on different cell types. These devices are able to produce hundreds of thousands of images per day. One goal of the cell assay image analysis is to label a few selected cell images by hand and to automatically label the vast majority of the images afterward. To obtain a classification of one image, it is segmented into small subimages, each containing one cell of the original image. Segmentation allows the cells to be considered separately in order to distinguish between different reactions of cells in the same image. When most of the small subimages are classified, a classification of the original image can be made by a majority decision.

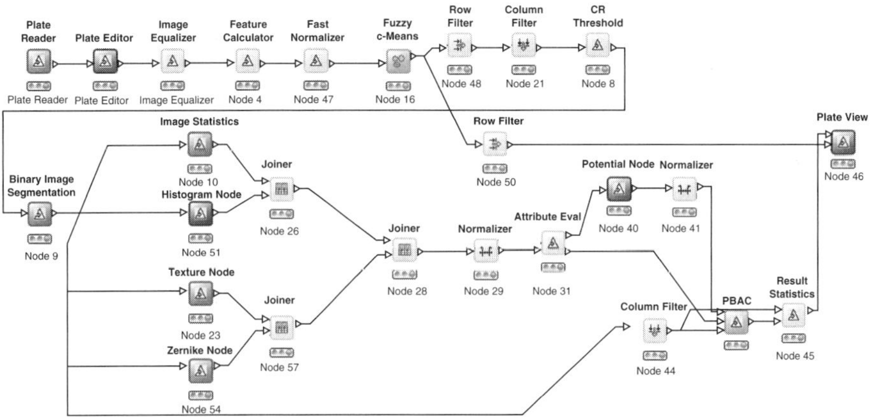

FIGURE 11 Workflow demonstrating the analysis of cell assay images by various image-processing nodes. (*See insert for color representation of the figure*.)

The workflow shown in Figure 11 consists of three major parts: the segmentation part, the feature extraction part, and the classification part. First, the raw images from the used plates are read in via the plate reader and the plate editor. Then the images are equalized, which means that their gray values are scaled to the complete range available. After that, several features such as histograms or textures are computed for each image as a whole. After normalizing the images' histograms, they are roughly clustered; some clusters are removed afterward (e.g., if they are too noisy). Then the images are segmented into small subimages, each containing a single cell. Obviously, the resulting number of data points is very large, because the thousands of images are segmented into small subimages. This results in an order of millions of images. The next nodes in the flow compute such features as texture, Zernike shapes, and peaks for each of the small single-cell images, which later are joined together in one big table. The last part of the workflow consists of the classification of the images, which in this case is performed using an *active learning* approach.[10] Briefly, this means that the user has to classify some (carefully selected) images first and then the algorithm only asks the user for images about which it is uncertain. The final classification of the complete images is obtained by classifying each individual cell within the given image. Each cell is assigned to a cluster and its corresponding class. The proportion of the distribution of the various classes is the decisive factor for classifying the entire image. If a clear majority decision can be made, the image does not need to be considered further. Figure 12 shows the plate view, in which the wells are assigned different classes indicated by the colors.

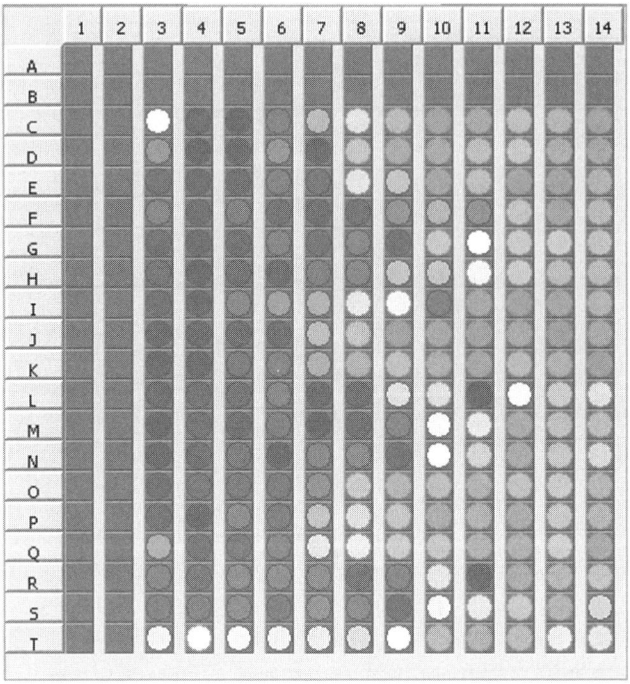

FIGURE 12 The plate viewer provides a graphical view of the classification results. Each point is a separate well on the plate, with the color indicating the class of the cell culture contained.

Text Mining in Scientific Publications

Text mining is a research area where unstructured texts are analyzed automatically and transformed into data structures that can be used with classical data-mining techniques such as clustering or even classification. Problems arising in text mining and natural language processing (NLP) are manifold: for example, part-of-speech tagging, stemming, named entity recognition (i.e., assigning words to different categories, such as gene or protein), and concept or keyword extraction. There is an extension for KNIME, which enables analysis of text documents, that contains special data types, such as documents, terms, sentences, and tags, together with a bunch of nodes. Figure 13 shows a big flow that processes PubMed abstracts for later clustering and classification. The two nodes at the top left take a simple query string and fetch the matching abstracts from PubMed. In the example two different document categories are retrieved, one dealing with "Human" and "Aids," the other with "Mouse" and "Cancer." After concatenating the two tables, which contain each matching document as a document cell in a row, first biomedical entities such as

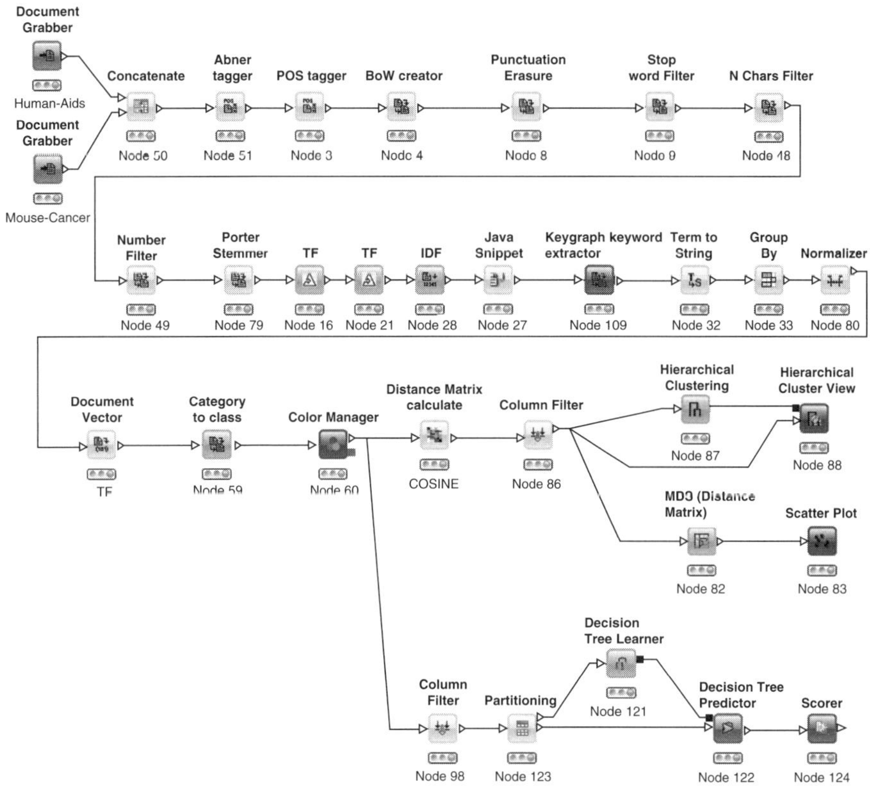

FIGURE 13 KNIME workflow showing the use of the text-mining extension. After several preprocessing steps, the documents are clustered and a decision tree for later classification of new documents is built.

"gene," "protein," or "cell line" are recognized using ABNER,[28] and then parts-of-speech (verb, noun, etc.) are assigned. Next a "bag of words" is created (i.e., tuples of single words and documents). The next four nodes remove terms that are generally useless for further analysis, such as stop words ("and," "is," "the," ...), punctuation characters, words that are too short, or numbers. The Porter stemmer then stems all terms and outputs only their word roots, which usually reduces the number of terms. Then relative and absolute frequencies for each term in each document are computed together with the inverse document frequency. The Java snippet node computes the *TF*IDF* value, which is the relative term frequency multiplied by its inverse document frequency. Next, the important keywords are extracted from the documents, in this case using a graph-based approach.[23] The following nodes change the table format so that it contains a document vector for each document, where the entries in the

Boolean vector represent the extracted keywords and indicate whether or not a word appears in the document. The category to class node adds to each row (either "Human-Aids" or "Mouse-Cancer") a column containing the class of the document, and the color manager adds color information based on the class assigned to each row (the colors can be chosen in the dialog). The last part of the workflow is to demonstrate two possible ways to build models from these document vectors. In the two upper branches, unsupervised models are built—a hierarchical cluster model and a 2D projection with MDS—using a precomputed distance matrix. For high-dimensional bit vectors such as the document vectors, it is preferable, for example, to use the cosine distance instead of the standard Euclidean distance. The two models can be visualized later. By using the appended color information, it is easy to determine the quality of the model just by looking at it. In the cluster dendrogram, branches should consist mainly of items from the same class at the leaves, and in the MDS, plot points with equal colors and shape (= class) should group together. This works quite well for the example data used, as can be seen in Figure 14 (we do not show the cluster dendrogram, as the different classes are only visualized

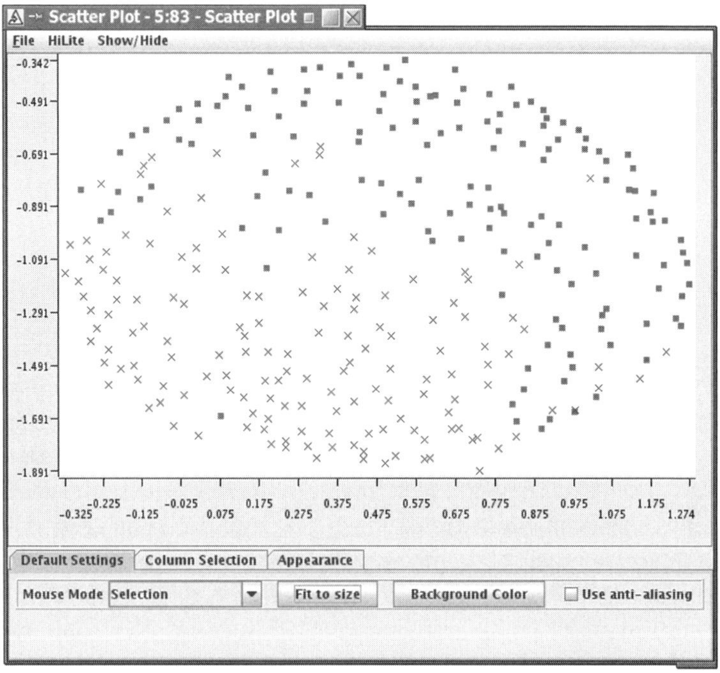

FIGURE 14 Two-dimensional projection of the document vector space from the MDS algorithm. The two document classes (squares and crosses) separate quite well.

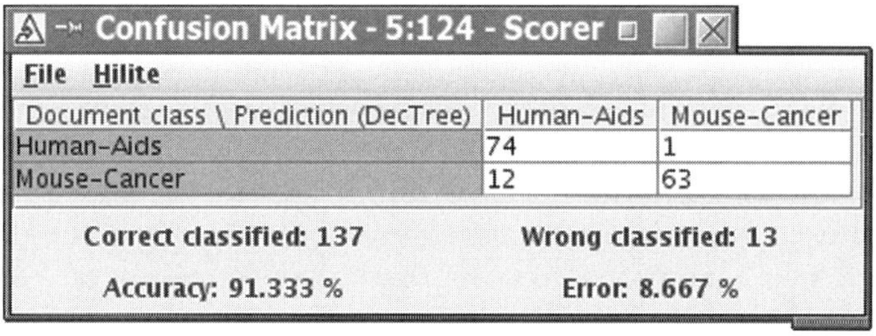

FIGURE 15 The scorer shows the confusion matrix of the prediction and the classification error and accuracy.

by small color points and cannot be distinguished well if printed in this format).

The lower branch performs supervised training of a decision tree by using the available category information. The built tree can be used later to classify unknown documents. In the example this is simulated by splitting the entire document table into two parts, one for training the other for testing. The scorer node at the end shows a confusion matrix and the error rate of the prediction (see Figure 15).

Acknowledgments

We want to thank the members in our group who have provided valuable comments while preparing this chapter (besides having implemented various parts of KNIME), especially Kilian Thiel for providing the text-mining flow and Nicolas Cebron for the cell assay image analysis example. Special thanks goes to Nycomed for the vHTS and cell image data used in the example flows. Last but not least, Heather Fyson did a great job of improving readability and spelling.

REFERENCES

1. Badreddin Abolmaali, S. F.; Wegner, J. K.; Zell, A. The compressed feature matrix: a fast method for feature based substructure search. *J. Mol. Model.*, **2003**, 9(4), 471–490.
2. Accelrys. Pipeline Pilot. http://accelrys.com/products/scitegic/.
3. Altintas, I.; Berkley, C.; Jaeger, E.; Jones, M.; Ludäscher, M.; Mock, S. Kepler: an extensible system for design and execution of scientific workflows. In *Proceedings of the 16th International Conference on Scientific and Statistical Database Management,* **2004**, pp. 423–424.

4. Berthold, M. R.; Cebron, N.; Dill, F.; Di Fatta, G.; Gabriel, T. R.; Georg, F.; Meinl, T.; Ohl, P.; Sieb, C.; Wiswedel, B. KNIME: the Konstanz information miner. In *Proceedings of the 4th Annual Industrial Simulation Conference, Workshop on Multi-agent Systems and Simulation,* 2006.
5. BIRT. Business Intelligence and Reporting Tools. http://www.eclipse.org/birt/.
6. Borgelt, C.; On canonical forms for frequent graph mining. In *Workshop on Mining Graphs, Trees, and Sequences at PKKD 2005*, pp. 1–12.
7. Borgelt, C.; Berthold, M. R. Mining molecular fragments: finding relevant substructures of molecules. In *Proceedings of the IEEE International Conference on Data Mining*. IEEE Press, Piscataway, NJ, 2002, pp. 51–58.
8. Brezany, P.; Janciak, I.; Min Tjoa, A. GridMiner: an advanced support for e-science analytics. In *Data Mining Techniques in Grid Computing Environments*. Wiley, Chichester, UK, 2008, pp. 37–56.
9. Chemistry Development Kit. http://cdk.sf.net/.
10. Cebron, N.; Berthold, M. R. Adaptive active classication of cell assay images. In *Knowledge Discovery in Databases: PKDD 2006* (*PKDD/ECML*), Vol. 4213. Springer-Verlag, Berlin, 2006, pp. 79–90.
11. ChemAxon. http://www.chemaxon.com/.
12. DataMining Grid. http://www.dataminininggrid.org/.
13. The Eclipse Project. http://www.eclipse.org/.
14. Nycomed Chair for Bioinformatics and Information Mining at the University of Konstanz. KNIME: Konstanz Information Miner. http://www.knime.org/.
15. Infocom Corporation. http://www.infocom.co.jp/index_e.html.
16. InforSense. InforSense KDE. http://www.inforsense.com/kde.html.
17. Insightful. Insightful Miner. http://www.insightful.com/products/iminer/default.asp.
18. Kepler Project. http://www.kepler-project.org/.
19. UC Los Angeles Laboratory of Neuro Imaging. LONI Pipeline. http://pipeline.loni.ucla.edu/.
20. Molecular Discovery Ltd. http://www.moldiscovery.com/.
21. myGrid. http://www.mygrid.org.uk/.
22. myGrid Project. Taverna Project Website. http://taverna.sourceforge.net/.
23. Ohsawa, Y.; Benson, N. E.; Yachida, M. KeyGraph: automatic indexing by co-occurrence graph based on building construction metaphor. In *ADL '98: Proceedings of the Advances in Digital Libraries Conference*. IEEE Computer Society, Washington, DC, 1998, pp. 12–18.
24. Oinn, T.; Greenwood, M.; Addis, M.; Alpdemir, N. M.; Ferris, J.; Glover, K.; Goble, C.; Goderis, A.; Hull, D.; Marvin, D.; et al. Taverna: lessons in creating a workflow environment for the life sciences. *Concurrency Comput. Pract. Exper*. **2006**, 18(10), 1067–1100.
25. R Project. The R Project for Statistical Computing. http://www.r-project.org/.

26. Rarey, M.; Dixon, J. S. Feature trees: a new molecular similarity measure based on tree matching. *J. Comput.-Aided Mol. Des*. **1998**, 12(5), 471–490.
27. Schrödinger, Inc. http://www.schrodinger.com/.
28. Settles, B. ABNER: an open source tool for automatically tagging genes, proteins, and other entity names in text. *Bioinformatics* **2005**, 21(14), 3191–3192.
29. Sieb, C.; Meinl, T.; Berthold, M. R. Parallel and distributed data pipelining with KNIME. *Mediterr. J. Comput. Networks* **2007**, 3(2), 43–51.
30. Steinbeck, C.; Hoppe, C.; Kuhn, S.;, Floris, M.; Guha, R.; Willighagen, E. L. Recent developments of the chemistry development kit (CDK): an open-source Java library for chemo- and bioinformatics. *Curr. Pharm. Des*. **2006**, 12(17), 2111–2120.
31. Symyx. http://www.symyx.com/.
32. Treweren Consultants. http://www.treweren.com/.
33. Tripos, L. P. http://www.tripos.com/.
34. Cardiff University. Triana. http://www.trianacode.org/.
35. Witten, I. H.; Frank, E. *Data Mining: Practical Machine Learning Tools and Techniques*, 2nd ed. Morgan Kaufmann, San Francisco, 2005.
36. Zauhar, R. J.; Moyna, G.; Tian, L. F.; Li, Z. J.; Welsh, W. J. Shape Signatures: a new approach to computer-aided ligand- and receptor-based drug design. *J. Med. Chem*. **2003**, 46(26), 5674–5690.

CHAPTER 8

USING CHEMICAL STRUCTURE TO INFER BIOLOGICAL FUNCTION

ANGELO D. FAVIA
Drug Discovery and Development, Istituto Italiano di Tecnologia, Genoa, Italy

IRENE NOBELI
Institute of Structural and Molecular Biology, Department of Biological Sciences, Birkbeck, University of London, London, UK

1 INTRODUCTION

The data explosion experienced in biology in the last couple of decades as a direct result of advances in genomic sequencing and macromolecular structure determination has inevitably left a gap between the number of proteins we know of and the proportion of these proteins whose function we know. The large number of orphan sequences and possible novel functions have created the need for automated, high-throughput approaches to function identification. Much of the recent progress in protein function identification has been documented extensively in excellent reviews of the subject.[1–8]

As biology raced ahead, it was not long before it was realised that chemistry could play a significant role in the quest for function identification, and many experimental approaches relying on essentially chemical methods have been devised to profile the function or functional state of a protein. In activity-based protein profiling (ABPP), chemical probes are directed at the active site of proteins, with the aim of determining their functional state, by exploiting the high specificity of the inhibitor structure for the active form of the protein in question. Labeling enzymes with small-molecule probes, such as tagged irreversible inhibitors, has been

Computational Approaches in Cheminformatics and Bioinformatics, First Edition.
Edited by Rajarshi Guha and Andreas Bender.

employed, for example, to distinguish active from inactive enzymes, as in the case of serine hydrolases tagged with biotynilated fluorophosphonates[9] or glycosidases targeted by probes containing glucose as the recognition head and biotin as the reporter group.[10] In fact, such probes have been generated for more than 20 enzyme classes.[1] The presence of large reporter tags in ABPP methods hinders the uptake and distribution of these molecules in vivo, thus limiting their applicability in profiling activity in a living cell, and alternatives have been sought that employ click chemistry to attach the tags in vitro, following the binding of the tag-free probe to the enzyme in vivo.[11] The idea of using biologically active small molecules as molecular probes is also at the core of chemical genetics and has been used for the characterization of proteins, whose function is difficult to decipher with genomic approaches such as gene knockouts. The method, pioneered by Shokat and co-workers,[12–14] has been used successfully to unravel the substrate specificity of selected kinases. The conserved ATP-binding site of protein kinases is engineered to recognize an ATP analog that is not recognized by the wild-type proteins, while at the same time the substrate specificity of the kinase itself is not affected. Often, engineering the mutant can be achieved with a single amino acid residue change involving a bulky "gatekeeper" residue being changed into a smaller amino acid that creates a hole unique to the mutant protein, and absent from the native. In this way, the mutant proteins can be inhibited with high specificity, and, importantly, within a time scale that permits inhibition to reveal the protein's native function. Finally, metabolomics methods are another example where chemical structure is used (albeit indirectly) to unravel protein function. Targeted and untargeted (discovery) metabolite profiling methods essentially monitor the presence and quantities of small-molecule natural products, thus illuminating the consequences of disrupting metabolic networks, and highlighting the biochemical and physiological role of knocked-out or inhibited proteins in the context of the cell.[1]

These are only a few of the many examples of using small-molecule probes to unravel protein function in experimental studies. However, the cost of real experiments in both time and money, and the many practical challenges associated with the handling of several proteins, have driven the need for the development of computational approaches that can help guide experiments and reduce cost. Most of these approaches rely, one way or another, on chemical structure to infer function, and often have their roots in traditional computational chemistry or chemoinformatics approaches. The term *chemical structure* can obviously refer to the structure of the macromolecule whose function we aim to identify, but here we are interested in the structure of other molecules (usually of relatively small size) that help us define function based on their property of binding the protein of interest.

Significant advances in chemoinformatics have led to the development of a number of successful computational approaches for relating small molecules to their targets (usually in the context of finding inhibitors for enzymes, receptors, etc.), and most of these are also applicable to the reverse problem of finding the natural partner of a protein, thus elucidating its function. Jenkins et al.[15] have categorized these approaches into four methodology groups. The first approach, chemical similarity searching, relies on the premise that similar molecules are likely to share the same targets, and vice versa.[16,17] Methods in this group are relatively simple and easy to implement, and are generally fast, so are suitable for large-scale studies. They exploit the large number of technologies available for comparing small molecules, including two- and three-dimensional descriptors, structural keys, physicochemical, topological, and field-based descriptors, and so on.[18,19] Whereas in target identification one would simply search through a database of known ligand–target pairs with the structure and properties of a new ligand whose target is unknown, in function identification one could use a known synthetic ligand to try and guess the natural counterpart. Alternatively, the similarity of the targets could be used to identify targets whose ligands are known, and these ligands could then be used as clues to the nature of the cognate partner for the remaining targets. The second group of approaches involves data mining and machine learning. Here again, a large number of methods that have been borrowed from computer science and statistics, or even developed especially for chemoinformatics applications, are now available (see, e.g., relevant chapters in Gasteiger and Engel,[20] Leach and Gillet,[21] and Bajorath[22]).

The third group of approaches involves bioactivity spectra (i.e., collections of activities or effects of a compound against a panel of proteins, a series of cells, a DNA microarray, etc.[23,24] Reversing this methodology, for example, by building a panel of activities of a number of selected "reference" compounds against a single protein, would also result in a *fingerprint* that could be used to associate proteins to functions. Finally, the fourth group of computational approaches relates to docking methods, especially panel docking, where either a single ligand is docked against multiple proteins[25] or multiple ligands are docked against a single protein, the latter being a way of identifying natural partners of proteins, thus unravelling their function. Cross-docking multiple proteins and multiple substrates could also be used, as creating a "background" distribution of energies relating to both the binding site and the candidate substrate could help assign a statistical significance to the estimated binding energy for a particular pair.

In this chapter we concentrate on this last type of method (i.e., applications of molecular docking to assist protein function prediction). As this is a relatively new application for docking, there have been few directly relevant studies. We focus here on the limited suggested approaches in these published studies, and back up and enrich our review and critique of their strengths and caveats with the vast existing literature on other applications of docking. For the benefit of the uninitiated reader, we start with a short introduction to the theory of traditional molecular docking.

2 OVERVIEW OF TRADITIONAL PROTEIN–SMALL MOLECULE DOCKING: THEORETICAL BACKGROUND

The optimization problem of determining the best alignment of two molecular entities with respect to each other is commonly referred to as *molecular docking*. Given the three-dimensional (3D) structure of a protein (determined experimentally, or modeled computationally), compounds can be docked into its surface clefts, with the aim of identifying energetically favorable poses. Since protein inhibition is an important mechanism through which therapeutic effects can be achieved, docking has been a very popular approach to drug design and lead optimization. In addition, in the last decade, mostly thanks to the increasing number of protein structures with unknown biological function, docking simulations are being suggested as tools to predict the naturally occurring partners of enzymes (i.e., substrates).

A docking simulation is usually composed of two separate but linked steps: the search process and the estimation of the binding energy. The idea is to generate multiple poses during the search phase and then associate each one with an estimated binding energy. The first issue in docking is generally the definition of the area to search. Even though for over 80% of proteins the ligand binding site has been shown to coincide with the largest surface cleft,[26] the accurate detection of a ligand binding site, especially in the case of very large clefts, remains a challenge. Useful insights can be provided from the study of a member of the same family with a known mechanism of action, and from the residue conservation analysis across the family. The presence in the X-ray structure of a ligand bound is also a good common starting point for defining the area of interest.

Once the search area has been determined roughly, affinity maps (usually one for each atom type present in the ligand to be docked) are calculated. The interaction sampling is usually done setting a grid center

(center of mass of key residues, or ligand bound, when available), a grid spacing (typically, fractions of Ångstrom), and a number of points in each of the three directions (x, y, z). Pairwise interaction values (between protein atoms and a probe) are then calculated at each point of the grid. This step allows for significant savings of computational resources and speeding up of the protocol, since interaction values are stored for each atom type, and for each grid point in space, and they can be reused for all matching atom types and any new ligand containing them.

Determining the location of a rigid body within a static three-dimensional space is the solution to a mathematical problem that has six degrees of freedom, three translational and three rotational (see Figure 1). With few exceptions, molecules, cannot be treated as rigid bodies; hence, the degrees of freedom of docking a flexible ligand into the active site of a rigid protein increase with the number of torsional angles in the small molecule. Since a systematic sampling is often too demanding (even if several poses are a priori usually rejected due to steric clashes with the protein atoms), genetic algorithms, anchor and growth methods, and Monte Carlo samplings (often in conjunction with simulated annealing), among other methods, are employed to solve the problem. In a standard docking simulation, no bonds are formed or broken and usually polarisation effects are neglected (i.e. the charge distributions of the interacting molecules do not vary during the simulation time). At this stage each pose generated is associated with an estimated binding energy that depends on the chosen scoring function.

A standard molecular mechanics-based scoring function is usually composed of several modular terms that take into account nonbonded interactions as well as internal energy terms, and it is usually parameterized against experimental data. Equations vary widely among different scoring functions, but some terms are commonly present, most notably the van der Waals, and the hydrogen bond energy contributions (with the latter usually being treated as a particular case of van der Waals interaction), the ionic interactions, the loss of ligand internal degrees of freedom, and a desolvation term:

$$\Delta G_{\text{bind}} = \Delta G_{\text{vdW}} + \Delta G_{\text{Hb}} + \Delta G_{\text{elec}} + \Delta G_{\text{rot}} + \Delta G_{\text{sol}}$$

Individual terms are often multiplied by a penalty function that accounts for deviations from ideal geometric values of distance and angle. Additional terms may be present, such as the lipophilic contribution, which is proportional to the lipophilic contact surface between the protein and the ligand.

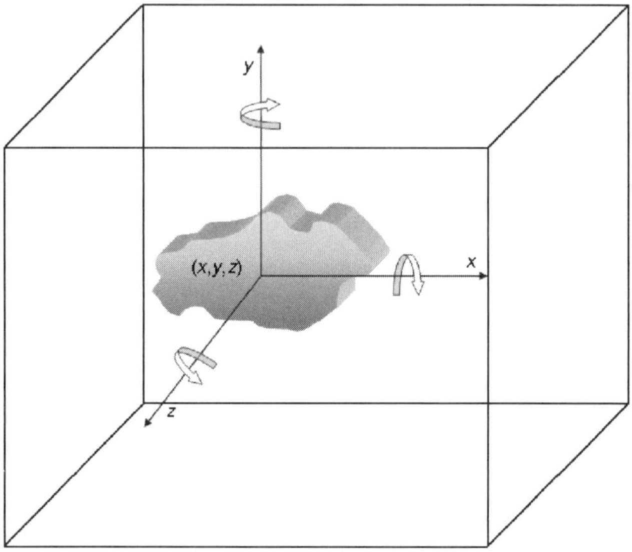

FIGURE 1 Absolute orientation of an object within a 3D space. There are six degrees of freedom: three translational (x, y, z coordinates of the center of mass) and three rotational, one around each of the axes.

The presence or absence of energy terms, and the different parameterization procedures result in scoring functions that vary widely in the reward or penalty terms for the same docking poses. The empirical nature of current scoring functions means that none of them is *ideal* and, under different conditions, protocols can, in turn, perform better than the others. Consensus scoring (i.e., combining results from several scoring functions) has been suggested as a possible solution and is usually shown to be superior to using the individual scoring functions on their own.[27–31]

Ideally, a docking run could be terminated once the lowest energy (global) minimum has been found, after exhaustive sampling of the available 3D space. However, exhaustive sampling is usually not feasible, and thus it is unlikely that a docking run will produce one unique solution. Users have to expect several diverse poses as the output of a docking run, all of which are, to some extent, *correct* solutions (i.e., local minima), ranked according to their estimated energy values.

3 DOCKING FOR PROTEIN FUNCTION IDENTIFICATION

As we mentioned already, docking has relatively recently found a new application in protein function identification.[32–41] For nonenzymatic proteins that employ small molecules to perform their biochemical and physiological role, it is obvious how identifying an energetically favorable interaction with a small molecule might give hints toward their function. In the case of enzymes, a substrate must first approach the catalytic residues and then establish relatively weak nonbonded interactions within the cleft. Here again, naturally interacting (cognate) substrates and enzymes must show adequate complementarity before the actual chemical transformations take place (although clearly this complementarity does not need to preexist for either the ligand or the enzyme).

The docking procedure in this case is similar to screening for inhibitors, but a number of additional issues arise that make this application more difficult in many ways. From the choice of the molecules to be docked, to the fact that many substrates exhibit only a weak to medium affinity for their cognate enzymes, the caveats in docking for function identification are numerous. In the next sections we discuss the most pressing challenges and possible ways of overcoming them.

Selecting What to Dock

The first issue in docking for function identification is selecting a panel of potential candidates for in silico screening against the protein of interest. The choice of this panel is far from trivial. Enzymes may act on a theoretically unlimited number of substrates, but in practice their cognate partners are usually limited by evolution to a very restricted set of structures encountered in nature. For example, although chemical databases such as the Available Chemicals Directory (http://accelrys.com) or PubChem[42] may contain millions of small molecules, the KEGG ligand database,[43] representing primarily molecules interacting with enzymes, currently counts just over 16,000 molecules. Although this is certainly an underestimation of the total number of naturally occurring molecules (especially as most natural products from plants or microorganisms with extensive secondary metabolism are not included), it is indicative of the difference in scale between data sets suitable for inhibitor screening and those suitable for function identification. The smaller number of possible cognate substrates should, in theory, make the process of docking easier. However, the question of whether existing databases cover adequately the spectrum of substrate structures for novel enzyme functions is debatable.

The most obvious solution to the problem is to dock as many compounds as possible, in the somehow optimistic hope that the cognate partner of the protein lies among them. There are a number of databases on the Internet that contain small molecules, such as KEGG,[44] ChEBI,[45] PubChem,[42] ChemBank,[46] and recently, efforts have also been made to store molecules in a format ready for docking (ZINC[47]), with some including multiple conformers for each ligand (CoCoCo[48]). As docking methods have become very fast, it is not unusual now to dock a few million compounds to a single receptor. However, this "brute force" approach has its disadvantages: It does not allow for more accurate scoring approaches to be implemented, and it may be less useful in cases where the substrate is not actually part of the database. Alternative approaches have been suggested, especially in the context of in silico screening for inhibitors, but these are obviously applicable to screening for substrates. One such approach is to avoid including members of the same ligand family in the docking hit lists. Su et al.[49] enriched the diversity of their hit list of candidate inhibitors for thymidylate synthase, dihydrofolate reductase, and a mutant lysozyme, by including in their final ranking only one representative from each structural family of compounds in their data set. In this approach all compounds from the Available Chemicals Directory were docked, but only family representatives (selected by clustering molecules based on their largest rigid fragments) were ranked, allowing for a more diverse list of hits. Alternatively, one can restrict the number of compounds selected for docking by exploiting knowledge of known ligands for the given protein, or of its mechanism of action. For example, Rastelli et al.[50] have used 3D pharmacophores derived from the hydrogen-bonding ability and volume templates of known inhibitors of dihydrofolate reductase to focus their screening data set on compounds with chemical features deemed necessary for inhibition of this protein. Song et al.[39] restricted their libraries of ligands to L,L-dipeptides and *N*-succinyl amino acids, as these are known substrates for the reaction catalyzed by the family of proteins to which the enzyme they were studying belonged. Similarly, Rakus et al.[38] used a focused library of limited chemical structures (such as monoacid and diacid sugars) to guide experiments for function prediction of an uncharacterized enolase, after inspecting the results of a wider docking screen using the whole of KEGG. In a more elaborate approach, Hermann et al.[33] created a filtered subset from KEGG by allowing ligands to be screened only if they contained an electrophilic substructure required in the catalytic mechanism of the family of amidohydrolases they were studying. In general, conserved substrate substructures within enzyme superfamilies, such as those discovered by Chiang et al.,[51] can be used to build focused libraries of relevant substrates and intermediates by filtering a priori large data sets with predefined chemical/functional groups.

We believe that a more satisfying approach to docking vast databases of small molecules may be to explore the relevant chemical space as well as possible by using a diverse (but limited in number) range of compounds. In our own exploration of substrate specificity in the short-chain dehydrogenase/reductase (SDR) family of proteins,[32] we opted for a relatively small but structurally diverse subset of the human metabolome. Human metabolites from KEGG were clustered using hashed fingerprints, and the "centroid" representative of each cluster was used for docking, resulting in only 115 candidate molecules for docking. We found that this was a reasonable approach for structurally coherent clusters and for highlighting classes of compounds that in the absence of any other knowledge may be better binders for a given active site. This approach is also backed up by studies of crystal structures from the PDB which show structurally similar ligand binding to the same protein in the same orientation, albeit often accompanied by side-chain movements and rearrangement of tightly bound waters.[52] However, it is important to recognize that true positive candidate substrates can be missed in this way, especially if none of the cluster representatives share a high degree of similarity to them.

Overall, we think that docking a large, comprehensive database that covers as many of the known substrates as possible is currently the most likely approach to lead to a few good candidates for experimental verification if standard high-throughput docking methods are applied. However, docking fewer and more diverse structures is likely to be more beneficial in cases where the conformational flexibility of the binding site dictates the use of time-consuming molecular dynamics approaches, or where the nature of the catalytic site might benefit from more accurate modeling, such as inclusion of quantum mechanical effects. In addition, a data set of diverse structures, identified computationally as useful for function identification, could be invaluable for in vitro (or even in vivo) screening, where using millions of compounds for screening is simply unrealistic.

Filtering Poses: Pre- and Postdocking

During the docking simulation, ligands are allowed to move within a user-predefined volume. Depending on the size and physicochemical nature of the search space, a docking run can produce, for a single ligand, several poses, corresponding to as many minima in the free-energy landscape. In traditional docking studies involving screening of many ligands, often only the top-scoring pose for each ligand is considered. However, this pose does not necessarily represent a productive spatial arrangement of a substrate with respect to an enzyme. In other words, energetic criteria are not enough to distinguish catalytically competent complexes from others, and additional geometric criteria must be taken into account. Good substrate candidates, for example, must bind in a geometry that allows catalytic residues to be in close contact with the ligand's reactive center. In this case, a requirement for filtering the poses is the knowledge of which residues are actually important for catalysis and which ligand atoms are involved in the process. Where this knowledge is not directly available, clues can be obtained by putting the studied protein in its wider family context. In the case of enzymes, the general consensus is that proteins of significant sequence identity (above 40%), when exhibiting functional variation, often share the same chemistry, but may act on diverse substrates. This means that the amino acids interacting directly with the substrate (i.e., that undergo some chemical transformations) should be conserved, while the ones that have an accessory role (i.e., recognition or complex stabilization) may vary. For example, esterases usually share a conserved catalytic triad (Ser–His–Glu/Asp) at the active site, probably because this is the most efficient way that proteins have evolved to split an ester into an alcohol and an acid. However, those enzymes have diverged, creating unique environments around the triad itself, in order to recognize and accommodate different esters.

Filtering using geometric criteria can take place either pre- or postdocking. In predocking filtering, geometric constraints are used to limit the acceptable ligand poses to a set that satisfies predefined criteria. Constraining, for example, a ligand atom to remain within a certain distance from a protein atom (usually because this pair of atoms is directly involved in catalysis) will restrict the search space and reduce the number of poses that need to be scored and ranked. This "anchoring" of the ligand to a specific residue or position in the catalytic site is a common and simple approach to ensure better-quality poses for substrate candidates, but in the absence of information on the ligand reaction center, and when dealing with larger ligands, such constraints may be too generic to guarantee catalytically competent complexes (e.g., a distance constraint of 3.0 Å from any ligand atom to a protein catalytic residue may not be very helpful if the ligand is over 15 Å in length and the part that is involved in catalysis is unknown a priori).[32] In an example of postdocking filtering, Tyagi and Pleiss[40] used geometric constraints to filter down their top 100 hits, including distances and angles relating to catalytic residues and the oxyanion hole and oxyanion hydrogens in alpha and beta hydrolases.

Fine-tuning the Energy Scoring Function and Rescoring Docked Poses

Scoring functions used in docking attempt to approximate the change in free energy occurring when the ligand binds to the protein. There are a number of methods that can estimate relative and absolute free energies of binding by computational means in a reliable way.[53,54] Unfortunately, these methods are too computationally expensive to be used in a virtual screening simulation, where millions of poses need to be evaluated and ranked. For this reason, alternative approaches are commonly used. Current docking programs use a variety of scoring functions to estimate binding affinity, including force-field-based, knowledge-based, and empirical scoring functions. Scoring functions, in general, consider the free energy of binding as a sum of additive terms, each of which reflects a particular contribution to the binding. Usually, tunable parameters are then adjusted to fit a set of experimental data. In these algorithms, for reasons of computational efficiency, some energy terms are commonly poorly treated.

Despite their obvious limitations, commonly employed scoring functions have had considerable success not only in traditional docking, but also in docking for function identification. For example, screening metalloenzymes for substrates (and inhibitors)[35] showed that even in the presence of metals, a classical scoring function and traditional docking methodology may be adequate. The only deviation from standard procedures in this case was the assignment of formal charges to the metal and the protein ligating residues to reflect expected charge transfer from the metal to these residues.

A number of reviews have attempted to compare different docking programs and the corresponding scoring functions (usually, in the context of screening for inhibitors).[55–60] Two lessons learned from such comparisons are that (1) no scoring function consistently outperforms all others given a wide range of binding sites and ligands (although some functions are consistently worse than others), and (2) most scoring functions perform at best adequately when ranking the relative binding affinity of multiple ligands to the same receptor, but fail the test of accurate prediction of the changes in free energy upon binding. In fact, it is widely accepted that the estimated energies for a wide range of protein–ligand complexes do not correlate significantly with experimentally measured binding affinities.[60,61]

A crucial role in the binding process is played by the solvent. In fact, in order to bind, ligands must displace water molecules from the binding cleft. Standard docking calculations usually treat solvent implicitly and cannot account properly for the energetic term associated with its displacement. More sophisticated algorithms try to predict the location of water molecules within the active site, but even these are far from modeling the solvent effects properly.[62–65]

An important role in the binding process is played by electrostatic forces, which remain nonnegligible at relatively long interatomic distances, and whose strength depends on the dielectric constant of the medium separating the atoms involved, which is difficult to predict accurately. Electrostatic interactions are notoriously difficult to calculate accurately, and hence this term is often approximated poorly in docking simulations (occasionally, it is left out altogether, to speed up calculations). Polarization effects are totally neglected, too, even if, in reality, the charge distributions of two interacting molecular entities vary according to their respective spatial orientation. In standard docking, protein and ligand charge distributions do not change during the simulation time. All these approximations contribute to the large numbers of false positives and negatives observed in most docking calculations.

To overcome the shortcomings of state-of-the-art scoring functions, standard docking calculations are now often followed by rescoring of a limited number of poses, deemed to be most promising after the search. This approach is based on evidence that most docking programs are able to find reasonable protein–ligand poses (which result from comparing different poses of the *same* ligand) but do a poorer job when it comes to ranking poses of different ligands. Rescoring procedures, which are typically more accurate but also much slower than common scoring functions, aim to improve the ranking step without adding a significant time overhead to the search step.

Molecular mechanics/Poisson–Boltzmann surface area (MM–PBSA) and molecular mechanics/generalized Born surface area (MM–GBSA) are two common methodologies for rescoring that have been successful in identifying the true ligands for several enzymes.[66–70] These methodologies have the advantage of treating the electrostatics and solvation components of the binding energy (which are poorly treated in standard docking) with a higher level of theory. The binding energy (E_{bind}) is obtained by subtracting from the energy of the complex (E_{complex}) the energies of the ligand and the protein (E_{ligand} and E_{protein}, respectively). All terms are calculated taking into account the solvent effects:

$$E_{\text{bind}} = E_{\text{complex}} - E_{\text{ligand}} - E_{\text{protein}}$$

During the calculation of the E_{complex} term the ligand is commonly allowed to "relax" within the active site, while the protein is held fixed. An induced-fit effect can be simulated if the protein structure were also allowed to adapt upon binding. This would indeed resemble better what goes on at a molecular level, but the fact that no major structural rearrangements are allowed and the low *ratio* of the added benefits to the calculation time limit the procedure's applicability enormously.

Even though these approaches have been used successfully in some cases and are currently the most appealing alternatives to more reliable but slower treatment of interaction energies, they have their own challenges. They cannot entirely capture the complex protein–ligand interaction process, especially the dynamic recognition process that takes place before the final bound pose is assumed. Moreover, the force fields and solvation energies are usually parameterized against a given set of experimental data; hence their accuracy is likely to be limited to cases of similar chemical nature to the original training set. Perhaps, more intriguingly, the question of whether rescoring procedures are always beneficial for ranking the proteins' natural counterparts is currently an object of debate. A pose that was a minimum in one potential energy surface is not necessarily optimal in a surface defined differently, so rescoring poses without further energy minimization can be misleading. A further minimization step involving specific residues in the binding site may be a reasonable compromise, especially in cases where only a homology-built model of the receptor is available.[39] In addition, the benefits of rescoring may vary depending on the nature of the catalytic site and the intermolecular interactions stabilizing the complex.

In our study of candidate substrates for the SDR family of proteins, we observed that rescoring with MM-GBSA worked best (unsurprisingly) for complexes where polar interactions were important. In cases of hydrophobic interactions, we believe that the Glide scoring function (www.schrodinger.com) was adequate, probably because like many other scoring functions, it has been optimized to reproduce known protein–ligand complexes, which are primarily hydrophobic in nature (as are most synthetic ligands in the PDB). In a much more comprehensive study of the effect of MM-GBSA rescoring of docking hits to three buried model binding sites, Graves et al.[71] noted that rescoring rescued several false negatives of the standard docking procedure, but also introduced many false positives, especially at the top hit-list ranks. The same authors identified the relaxation of the complex allowed by MM-GBSA as the most important source of both the advantages and disadvantages of this method. In general, a more accurate function based on higher-level theory is what docking studies aspire to, but these functions could actually perform worse, simply because they cannot account for effects that are often included implicitly in simpler and less physically meaningful scoring functions.

Finally, it should be pointed out that in molecular mechanics–based simulations (such as docking), electrons are not taken into consideration, thus bonds are neither broken nor formed. As a consequence, docking algorithms, with or without rescoring procedures, aim *only* at reproducing poses that are likely to exist prior to the actual catalytic event. Whether a chemical transformation will effectively follow the binding process cannot be directly extrapolated from such a simulation, since the activation energy that determines whether a reaction is kinetically feasible is not considered

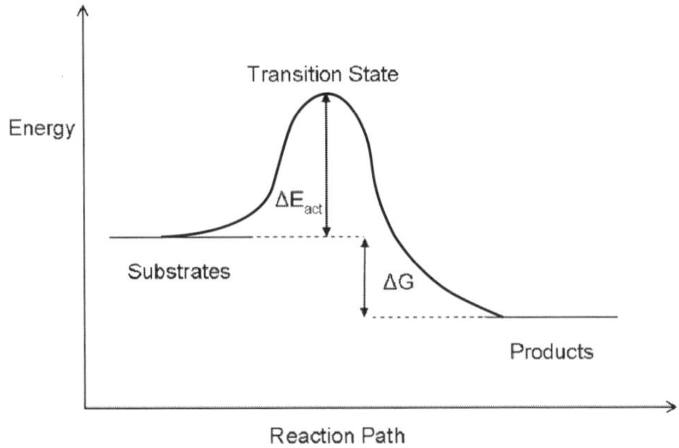

FIGURE 2 Simplified diagram of the change in the energy of a complex along the reaction coordinate. The "substrates" level represents the energy of the protein–ligand complex before the reaction takes place; the "products" level represents the stage at which the products are released. The transition-state step corresponds to the highest energy barrier along the reaction coordinate that needs to be overcome for the chemical transformation to take place.

(see Figure 2 for a simple schematic of the changes in energy involved in a generic chemical transformation as a function of the reaction path).

In principle, a way to deal with bond formation and breaking in order to better model catalytic events is the use of quantum mechanics–based methods. As it is impractical to apply such methods to large protein systems (due to the very high computational cost), hybrid quantum mechanics/molecular mechanics (QM/MM) approaches are now being employed in studies of individual enzymes and catalytic reactions.[72–74] These restrict the use of high-level theory to a limited number of residues in close contact with the substrate or otherwise important for catalysis, and assume standard MM energy functions for all other atoms. Their greatest challenge is dealing with the borders between the areas treated quantum mechanically and the rest of the system. Such methods are still too expensive to be applicable in the high-throughput screening required for function identification, but one could foresee that advances in applying parallel programming and grid technologies could make these methods more popular in the not too distant future.

Dealing with the Protein Conformational Flexibility

When docking programs were first developed, the available computational capabilities made even a simple task such as docking rigid ligands to rigid proteins a time-demanding issue. Later, following hardware and algorithmic advances, docking algorithms that took into account ligand flexibility became feasible and popular. Currently, allowing ligand flexibility is standard in docking protocols, and developers around the globe are focusing their efforts on efficient ways of treating protein flexibility instead.

Dealing with protein flexibility is an important issue; X-ray structures (the most commonly used structures in docking) capture only one of multiple low-energy conformations accessible by a given protein sequence. More specifically, when dealing with *holo* (ligand bound) structures, docking results can be favorably biased toward molecules resembling the ligand experimentally bound. On the other hand, *apo* (ligand-free) structures often have binding clefts that are too small to accommodate the ligand. Both issues arise because protein side chains adapt constantly to the presence or absence of bound ligands. An example of this adaptability is depicted in Figure 3, where the superposition of a set of experimentally determined structures of *Torpedo californica* acetylcholinesterase are shown. The single movement of just one amino acid (F330) allows the binding of very different inhibitors.

The scaling of van der Waals radii, and consequently the softening of the intermolecular repulsion between protein and ligand atoms at very short distances, is the cheapest and probably most commonly used technique by which docking algorithms can account for moderate induced-fit effects, without the computational cost of exploring different protein conformations. Another popular way of treating protein flexibility in docking simulations is to take into account more than one protein structure of the same enzyme. Depending on the conditions of crystallization, proteins may be captured at different local minima of the rugged energy landscape at the bottom of their folding funnel. Any of these structures may be better suited for binding the native substrate, so using as many of them as possible in docking increases the chances of finding a "holo-resembling" structure. For example, in the work of Tyagi and Pleiss,[40] 137 different crystal structures representing just 20 hydrolase complexes were used. In one example from this study, 12 substrates were docked to five different structures of *H. lanuginosa* lipase, and all of them (except a tertiary alcohol) could be docked to at least one of the structures, but for each substrate, at least one structure was not accessible. Ensembles of protein conformations produced by NMR structure determination experiments provide an alternative route to taking into account some flexibility, as they can be clustered, and representatives from each cluster can be used for virtual screening.

In an alternative approach that aims to capture the protein motions directly, docking is sometimes used in conjunction with molecular

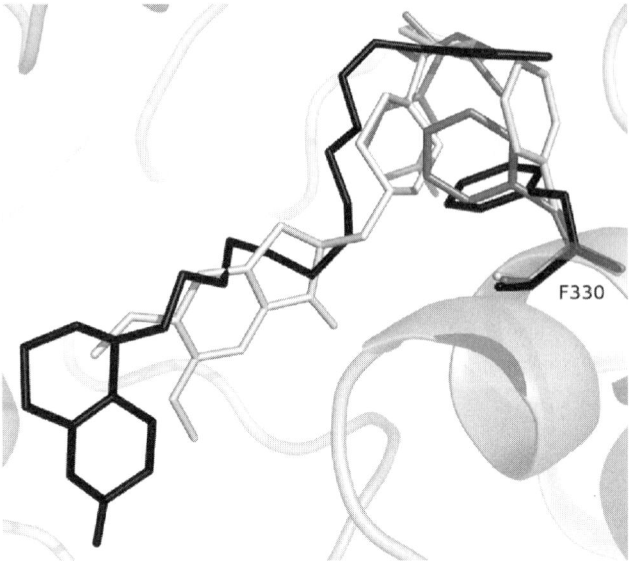

FIGURE 3 Structural superposition of three *Torpedo californica* acetylcholinesterase X-ray structures in complex with edrophonium, donepezil, and (*r*)-tacrine(10)-hupyridone (shown as stick models colored dark gray, light gray, and black, respectively). Phenylalanine 330 from the three structures is highlighted as a stick model, colored according to the inhibitor in the protein complex. The backbones and remaining side chains of the three proteins were perfectly superimposable. (Just one protein is shown in the figure, to help visualization.)

dynamics (MD) simulations (i.e., docking ligands to a number of snapshots taken from an MD trajectory). Computationally cheaper approaches such as normal mode analysis[75–77] or the CONCOORD[78] and tCONCOORD[79] distance constraint-based methods can potentially replace MD in sampling at least part of the conformational landscape of proteins prior to docking. All these approaches are limited by the computational penalty associated with the increase in the number of docking runs that must be performed, and by the fact that a suitable binding cleft for the native substrate may not exist among either the set of crystal structures considered or the snapshots resulting from MD simulations.

An alternative way to tackle the conformational flexibility problem is to split it into two separate parts: During the first step, ligands are docked to a protein structure whose relevant residues in the active site have been in silico mutated into less bulky ones (such as alanine).[80] In the second step, the original side chains are restored and relaxed around the docked ligands in order to establish favorable interactions. This protocol allows larger ligands to dock and enacts, to some extent, the protein–ligand mutual adaptation at short time scales but cannot reproduce slow, large-scale structural rearrangements such as loop or domain motions.

Docking High-Energy Intermediates

Perhaps the most promising approach to using small molecules for protein function identification involves the use of "high-energy intermediates" in docking. A series of papers[33,34,40] have recently described how docking metabolite structures that mimic the intermediates of plausible chemical reactions could be used to identify the catalytic function of an enzyme whose 3D structure has been determined. The use of intermediates is justified by the fact that enzymes must bind the transition state stronger than the substrate itself, or no increase in the catalytic rate would occur.[81] The authors of these papers have been very careful in highlighting many of the limitations of their method. As the use of intermediates for in silico function prediction seems to be gaining in popularity, we think it is worth taking a step back and considering both the limitations of this method, and, more important, why this method works.

When looking through a database of small molecules with the aim of identifying possible substrates, one clearly relies on the property of the enzyme being specific for its substrate(s). Specificity, as measured experimentally, would appear either in K_M or k_{cat}, depending on whether the affinity responsible for the increase in the reaction rate originates from contacts that are already present in the enzyme–substrate (ES) complex, or from new bonds formed when the substrate transforms to an excited (transition state) conformation.[82] Hence, when docking ground-state substrate candidates to enzymes, one assumes that specificity is reflected in K_M values, and by analogy, in the computationally predicted binding scores. Docking intermediate structures, on the other hand, is justified on the assumption that they are more likely to be discriminated than their ground-state equivalents, so the emphasis is on specificity reflected in the k_{cat} values (although obviously, specificity for the ground state is implicitly included for that part of the substrate structure that does not change).

The first limitation of this method is clear: Knowledge of the reaction mechanism is a prerequisite for producing the structures of the intermediates prior to docking (e.g., in the case of amidohydrolases, the knowledge that the transition state is a tetrahedral structure containing both the leaving and replacement groups is necessary to produce computationally likely candidates from ground-state structures). Hence, this method is not applicable where either no information is available about the overall function of the protein, or where the reaction mechanism is not conserved within the family or superfamily to which the protein belongs. In addition, the intermediate may not be an independent structure (as the one described above) suitable for docking studies. For example, in many catalytic reactions, the intermediate is covalently bound to a protein side chain that plays an active catalytic role (such as a nucleophile in protein abstraction from a substrate). Although Pleiss and co-workers[40,83] have been successful in reproducing such complexes in their docking studies, the majority of available docking programs are unable to handle covalent complexes. Even if this situation changes, short-term intermediates in catalytic reactions are likely to involve bonds with only partial covalent character to both substrate and protein residues, and as such, they are unlikely to be well modeled by conventional docking programs. In fact, in such cases it is more likely that quantum mechanical calculations (which are too time consuming for virtual screening) are necessary for successful modeling. Hence, as powerful as the use of intermediates may be, it cannot be a general method for function identification, given the structure of a protein alone.

Assuming that the intermediates of a reaction are known, there are additional problems that need to be addressed, most notably relating to why this approach, as it stands, should be successful. The question is not whether enzymes recognise their substrates better in their excited rather than their ground state (which they clearly do), but whether state-of-the-art docking programs are able to achieve the same discrimination, and if so, how. Results from recent studies strongly suggest that the answer to the first part of this question is yes. We believe that at least part of this success is due to the fact that enzymes examined so far are relatively easy cases, in that they contain metal ions crowding the catalytic site and imposing an electrostatic environment that decreases the likelihood of decoys with no perfect fit to the site to be accommodated. Although catalysis itself always requires a stringent geometrical alignment of substrate and enzyme features, promiscuous (but not necessarily reactive) binding of substrate analogs could be expected to a much greater extent in enzymes where the site is either larger, or with fewer prominent features, such as the metal ion. It would be interesting to see whether the docking of intermediates can be similarly successful in cases of enzymes with more forgiving binding sites (i.e., larger and less polarized clefts). Again, knowledge of the catalytic reaction may be crucial here, as it would guide modeling of the binding site (e.g., protonation states of catalytic residues) so that it accurately reflects the charge distribution experienced by a substrate entering the site.

Another important question that we think deserves answering is whether we can rationalize the differences in the docked scores between substrates in the ground and their excited states. To answer this question, we examine briefly three possible explanations that are invoked in the literature to rationalize the remarkable affinity of enzymes for transition states:

1. The enzyme conformation that binds the substrate in the ground state differs from that which interacts with the transition state and is achieved through dynamic stochastic searches.[84] The conformational changes that are either promoted or, more likely, stabilized by the transition state are often subtle,[85] but they are certainly contributing to the remarkable increase in binding affinity for the excited state. Observing a single enzyme molecule in action[86] has recently also highlighted the fact that the protein exists in a large number of conformers with different k_2 rates (in the Michaelis–Menten enzymatic reaction scheme). Hence, some of the conformations that are dynamically interchanging are a lot more catalytically competent than others, presumably because they achieve a more stable transition-state complex. The structures used in docking (which are often *apo* structures) have conformations unlikely to match the ones achieved during the binding of the transition state. Indeed, if the enzyme closes in on the transition state to achieve the extra affinity, it is difficult to justify how by using a highly populated conformation in the unbound state (as the one captured in the crystal structure and used in rigid-protein docking simulations) one could reproduce the marked differences between ground- and excited-state substrate binding.

2. Assuming that the actual conformation that interacts with the transition state is available, would the scoring function used in docking be able to tell the difference between a ground- and an excited-state substrate? Because cooperative effects are in action,[82] a difference in a single hydrogen bond, which would have only a modest effect in a docking score, could cooperatively achieve in reality a much larger effect during catalysis. Such effects cannot be reproduced by pairwise intermolecular potentials such as those used currently in docking studies.

3. Warshel and co-workers (see Warshel and Florián[87] and references therein) have put forward and tested in computer simulations over the years the suggestion that electrostatic stabilization of the transition state achieved by the preorganized dipolar environment of the catalytic site may be the most significant effect in catalysis. In contrast with a water environment, the enzyme does not need to pay the reorganization energy penalty associated with the formation of the transition state because this is already paid during the folding process. Although Warshel et al. proposed this as a mechanism for explaining the difference between the speed of reactions in water and within enzymatic active sites, one can deduce that if electrostatics play the biggest role in catalysis, the affinity for the transition state

must stem from strong electrostatic interactions. Hence, we can expect that reproducing successfully the electrostatic interactions between the protein and the substrate would reveal a large difference between the ground and the excited states of the ligand. The usual approximations in the force fields of docking studies are unlikely to be good enough, and undoubtedly part of the success of docking to amidohydrolases stems from the very careful treatment of electrostatic interactions.

If all the hurdles above exist, why has the method of docking intermediates been so successful? We already mentioned the possible contribution of a better treatment of electrostatics, including a careful atomic charge assignment that reflects known charge distributions among catalytic residues and substrates. The use of a low dielectric constant that corresponds to a hydrophobic binding site better matching the *holo* structure of the enzyme could also make up for the use of an *apo* structure. Finally, the presence of the charged metal atom in the cases studied must be, at least partly, responsible for the preorganization of the catalytic site, which, in turn, renders a tetrahedral ligand arrangement (such as the one in the intermediate state) much more favorable than the flat structure corresponding to the substrate in these studies.

Given the foregoing observations, one might not expect the docking of intermediates to work always as well as it has worked in the case of amidohydrolases. However, it is difficult to answer this question definitely without further exploring the use of intermediates with many diverse protein families and catalytic reactions. In addition, it is important that we understand where the success of this method stems from and how the limitations mentioned earlier can be overcome.

Other Issues

In previous sections we tried to address the major challenges in using docking for protein function identification. These were issues relating to the docking process itself, and one can imagine that at least some of them could be better addressed in the future following progress in docking and scoring methodologies, as well as advances in computer hardware. However, there also exist issues that may be harder to resolve, mostly because they are inherent to the problem of function identification in general.

We believe a major problem is the problem of interaction promiscuity.[88] Binding promiscuity among proteins and ligands is a well-established fact, and the basis for protein evolution, the phenomenon of compensating for a protein in a pathway after a gene knockout, and, of course, from a human point of view, drug development. Lately, ever-increasing experimental evidence suggests that functional promiscuity of proteins is also a lot more widespread than was thought initially.[89–93] It is clear that proteins are capable not only of binding multiple and often structurally unrelated ligands, but they may also possess flexibility in their catalytic machinery that allows them to accelerate reactions involving these ligands, albeit, of course, at different rates. Hence, searching for the natural (physiologically important) substrate of an enzyme becomes a rather more difficult task than that of identifying a ligand of relatively high affinity forming a catalytically competent complex with that enzyme. Even docking studies achieving very high ranks for the (known) substrates in in silico screens of thousands of metabolites still report a nonnegligible number of small molecules that rank better than the substrates in those screens.[37] It is at least possible that some of these molecules are indeed good binders of the proteins in question, although experimental validation would be needed to confirm this. To add to the difficulty of the task, it is very likely that, in vivo, the conditions in the cell, and more generally the environment of the protein of interest, play an instrumental role in its selectivity for its partners. For example, Saghatelian et al.[94] showed that substrate selectivity of the FAAH protein in vitro does not match the selectivity observed in vivo, and this, in our opinion, is likely to be a common problem. In such cases we must also expect that our in silico results cannot accurately reflect the in vivo results.

The problem of functional promiscuity reveals a broader issue in protein function identification using docking, or indeed any other method: the definition of function itself. The way we define function is directly related to our chances of success in predicting it. This was elegantly demonstrated by Babbitt[95] and others,[96] who pointed out that use of the current Enzyme Commission (EC) system for assigning function to enzymes is problematic for relating structure to function, and recommended the development of more refined systems that better reflect the mapping between structural and functional conservation within protein families. Other systems, such as Gene Ontology[97], may suffer from similar drawbacks, and one could imagine that perhaps superfamily-specific systems of functional annotation that reflect individual details of an evolutionary family might be better suited to function prediction methods that rely on the sequence and structure of the protein and its interactions with its partners.

4 CONCLUSIONS

In this chapter we described the state of the art in docking methodologies for protein function identification, focusing on the challenges that these methods face and discussing the solutions that have recently been suggested to overcome them. Although very few docking studies have been performed with the aim of predicting protein function, some important successes have been reported, and some genuine predictions for enzymes with unknown function have been made and have been verified with follow-up experiments (experimental verification is crucial given the alarmingly high levels of misannotation of protein function revealed in recent studies in some of the most popular databases). These promising early studies are opening the way for more research on the subject. This should lead to a better understanding of where the shortcomings are and to improvements in the methodologies. It is, however, clear that docking, as with any other computational approach, will need to be part of a larger effort involving as many diverse methods as possible if protein function prediction is to reap the rewards from the considerable efforts of computational biologists and chemists in recent decades.

REFERENCES

1. Saghatelian, A.; Cravatt, B. F. Assignment of protein function in the postgenomic era. *Nat. Chem. Biol*. **2005**, 1, 130–142.
2. Lee, D.; Redfern, O.; Orengo, C. Predicting protein function from sequence and structure. *Nat. Rev. Mol. Cell. Biol*. **2007**, 8, 995–1005.
3. Watson, J. D.; Laskowski, R. A.; Thornton, J. M. Predicting protein function from sequence and structural data. *Curr. Opin. Struct. Biol*. **2005**, 15, 275–284.
4. Godzik, A.; Jambon, M.; Friedberg, I. Computational protein function prediction: Are we making progress? *Cell. Mol. Life. Sci*. **2007**, 64, 2505–2511.
5. Hawkins, T.; Kihara, D. Function prediction of uncharacterized proteins. *J. Bioinf. Comput. Biol*. **2007**, 5, 1–30.
6. Friedberg, I. Automated protein function prediction: the genomic challenge. *Brief Bioinf*. **2006**, 7, 225–242.
7. Ofran, Y.; Punta, M.; Schneider, R.; Rost, B. Beyond annotation transfer by homology: novel protein-function prediction methods to assist drug discovery. *Drug Discov. Today* **2005**, 10, 1475–1482.
8. Frishman, D. Protein annotation at genomic scale: the current status. *Chem. Rev*. **2007**, 107, 3448–3466.
9. Liu, Y.; Patricelli, M. P.; Cravatt, B. F. Activity-based protein profiling: the serine hydrolases. *Proc. Natl. Acad. Sci. USA* **1999**, 96, 14694–14699.
10. Tsai, C. S.; Li, Y. K.; Lo, L. C. Design and synthesis of activity probes for glycosidases. *Org. Lett*. **2002**, 4, 3607–3610.
11. Speers, A. E.; Cravatt, B. F. Profiling enzyme activities in vivo using click chemistry methods. *Chem. Biol*. **2004**, 11, 535–46.

12. Shah, K.; Liu, Y.; Deirmengian, C.; Shokat, K. M. Engineering unnatural nucleotide specificity for Rous sarcoma virus tyrosine kinase to uniquely label its direct substrates. *Proc. Natl. Acad. Sci. USA* **1997**, 94, 3565–3570.

13. Liu, Y.; Shah, K.; Yang, F.; Witucki, L.; Shokat, K. M. Engineering Src family protein kinases with unnatural nucleotide specificity. *Chem. Biol*. **1998**, 5, 91–101.

14. Bishop, A. C.; Ubersax, J. A.; Petsch, D. T.; Matheos, D. P.; Gray, N. S.; Blethrow, J.; Shimizu, E.; Tsien, J. Z.; Schultz, P. G.; Rose, M. D.; et al. A chemical switch for inhibitor-sensitive alleles of any protein kinase. *Nature* **2000**, 407, 395–401.

15. Jenkins, J. L.; Bender, A.; Davies, J. W. In silico target fishing: predicting biological targets from chemical structure. *Drug Discov. Today Technol*. **2006**, 3, 413–421.

16. Mitchell, J. B. The relationship between the sequence identities of alpha helical proteins in the PDB and the molecular similarities of their ligands. *J. Chem. Inf. Comput. Sci*. **2001**, 41, 1617–1622.

17. Schuffenhauer, A.; Floersheim, P.; Acklin, P.; Jacoby, E. Similarity metrics for ligands reflecting the similarity of the target proteins. *J. Chem. Inf. Comput. Sci*. **2003**, 43, 391–405.

18. Bender, A.; Jenkins, J. L.; Li, Q.; Adams, S. E.; Cannon, E. O.; Glen, R. C.; David, C. S. Molecular similarity: advances in methods, applications and validations in virtual screening and QSAR. In *Annual Reports in Computational Chemistry*, Vol. 2. Elsevier, New York, 2006, pp. 141–168.

19. Willett, P.; Barnard, J. M.; Downs, G. M. Chemical similarity searching. *J. Chem. Inf. Comput. Sci*. **1998**, 38, 983–996.

20. Gasteiger, J.; Engel, T. D. *Chemoinformatics: A Textbook*. Wiley-VCH, Weinheim, Germany, 2003.

21. Leach, A. R.; Gillet, V. J. *An Introduction to Chemoinformatics*, Kluwer Academic, Dordrecht, The Netherlands, 2003.

22. Bajorath, J. *Chemoinformatics: Concepts, Methods, and Tools for Drug Discovery*. Methods in Molecular Biology, Vol. 275. Humana, Totowa, NJ, 2004.

23. Kauvar, L. M.; Higgins, D. L.; Villar, H. O.; Sportsman, J. R.; Engqvist-Goldstein, A.; Bukar, R.; Bauer, K. E.; Dilley, H.; Rocke, D. M. Predicting ligand binding to proteins by affinity fingerprinting. *Chem. Biol*. **1995**, 2, 107–118.

24. Lagunin, A.; Stepanchikova, A.; Filimonov, D.; Poroikov, V. PASS: prediction of activity spectra for biologically active substances. *Bioinformatics* **2000**, 16, 747–748.

25. Chen, Y. Z.; Zhi, D. G. Ligand–protein inverse docking and its potential use in the computer search of protein targets of a small molecule. *Proteins* **2001**, 43, 217–226.

26. Laskowski, R. A.; Luscombe, N. M.; Swindells, M. B.; Thornton, J. M. Protein clefts in molecular recognition and function. *Protein Sci*. **1996**, 5, 2438–2452.

27. Bissantz, C.; Folkers, G.; Rognan, D. Protein-based virtual screening of chemical databases: 1. Evaluation of different docking/scoring combinations. *J. Med. Chem*. **2000**, 43, 4759–4767.

28. Charifson, P. S.; Corkery, J. J.; Murcko, M. A.; Walters, W. P. Consensus scoring: a method for obtaining improved hit rates from docking databases of three-dimensional structures into proteins. *J. Med. Chem*. **1999**, 42, 5100–5109.

29. Gohlke, H.; Klebe, G. Statistical potentials and scoring functions applied to protein–ligand binding. *Curr. Opin. Struct. Biol*. **2001**, 11, 231–235.

30. Stahl, M.; Rarey, M. Detailed analysis of scoring functions for virtual screening. *J. Med. Chem*. **2001**, 44, 1035–1042.

31. Wang, R.; Wang, S. How does consensus scoring work for virtual library screening? An idealized computer experiment. *J. Chem. Inf. Comput. Sci*. **2001**, 41, 1422–1426.

32. Favia, A. D.; Nobeli, I.; Glaser, F.; Thornton, J. M. Molecular docking for substrate identification: the short-chain dehydrogenases/reductases. *J. Mol. Biol*. **2008**, 375, 855–874.

33. Hermann, J. C.; Ghanem, E.; Li, Y.; Raushel, F. M.; Irwin, J. J.; Shoichet, B. K. Predicting substrates by docking high-energy intermediates to enzyme structures. *J. Am. Chem. Soc*. **2006**, 128, 15882–15891.

34. Hermann, J. C.; Marti-Arbona, R.; Fedorov, A. A.; Fedorov, E.; Almo, S. C.; Shoichet, B. K.; Raushel, F. M. Structure-based activity prediction for an enzyme of unknown function. *Nature* **2007**, 448, 775–779.

35. Irwin, J. J.; Raushel, F. M.; Shoichet, B. K. Virtual screening against metalloenzymes for inhibitors and substrates. *Biochemistry* **2005**, 44, 12316–12328.

36. Kalyanaraman, C.; Imker, H. J.; Fedorov, A. A.; Fedorov, E. V.; Glasner, M. E.; Babbitt, P. C.; Almo, S. C.; Gerlt, J. A.; Jacobson, M. P. Discovery of a dipeptide epimerase enzymatic function guided by homology modeling and virtual screening. *Structure* **2008**, 16, 1668–1677.

37. Kalyanaraman, C.; Jacobson, M. P. Studying enzyme–substrate specificity in silico: a case study of the *Escherichia coli* glycolysis pathway. *Biochemistry* **2010**, 49, 4003–4005.

38. Rakus, J. F.; Kalyanaraman, C.; Fedorov, A. A.; Fedorov, E. V.; Mills-Groninger, F. P.; Toro, R.; Bonanno, J.; Bain, K.; Sauder, J. M.; Burley, S. K.; et al. Computation-facilitated assignment of the function in the enolase superfamily: a regiochemically distinct galactarate dehydratase from *Oceanobacillus iheyensis*. *Biochemistry* **2009**, 48, 11546–11558.

39. Song, L.; Kalyanaraman, C.; Fedorov, A. A.; Fedorov, E. V.; Glasner, M. E.; Brown, S.; Imker, H. J.; Babbitt, P. C.; Almo, S. C.; Jacobson, M. P.;

Gerlt, J. A. Prediction and assignment of function for a divergent *N*-succinyl amino acid racemase. *Nat. Chem. Biol*. **2007**, 3, 486–491.

40. Tyagi, S.; Pleiss, J. Biochemical profiling in silico: predicting substrate specificities of large enzyme families. *J. Biotechnol*. **2006**, 124, 108–116.
41. Macchiarulo, A.; Nobeli, I.; Thornton, J. M. Ligand selectivity and competition between enzymes in silico. *Nat. Biotechnol*. **2004**, 22, 1039–1045.
42. Wheeler, D. L.; Barrett, T.; Benson, D. A.; Bryant, S. H.; Canese, K.; Chetvernin, V.; Church, D. M.; DiCuccio, M.; Edgar, R.; Federhen, S.; et al. Database resources of the National Center for Biotechnology Information. *Nucleic Acids Res*. **2006**, 34, D173–D180.
43. Goto, S.; Okuno, Y.; Hattori, M.; Nishioka, T.; Kanehisa, M. LIGAND: database of chemical compounds and reactions in biological pathways. *Nucleic Acids Res*. **2002**, 30, 402–404.
44. Kanehisa, M.; Goto, S. KEGG: Kyoto encyclopedia of genes and genomes. *Nucleic Acids Res*. **2000**, 28, 27–30.
45. Brooksbank, C.; Cameron, G.; Thornton, J. The European Bioinformatics Institute's data resources: towards systems biology. *Nucleic Acids Res*. **2005**, 33, D46–D53.
46. Seiler, K. P.; George, G. A.; Happ, M. P.; Bodycombe, N. E.; Carrinski, H. A.; Norton, S.; Brudz, S.; Sullivan, J. P.; Muhlich, J.; Serrano, M.; et al. ChemBank: a small-molecule screening and cheminformatics resource database. *Nucleic Acids Res*. **2008**, 36, D351–D359.
47. Irwin, J. J; Shoichet, B. K. ZINC: a free database of commercially available compounds for virtual screening. *J. Chem. Inf. Model*. **2005**, 45, 177–182.
48. Del Rio, A.; Barbosa, A. J.; Caporuscio, F.; Mangiatordi, G. F. CoCoCo: a free suite of multiconformational chemical databases for high-throughput virtual screening purposes. *Mol. Biol.Syst*. **2010**, 6, 2122–2128.
49. Su, A. I.; Lorber, D. M.; Weston, G. S.; Baase, W. A.; Matthews, B. W.; Shoichet, B. K. Docking molecules by families to increase the diversity of hits in database screens: computational strategy and experimental evaluation. *Proteins* **2001**, 42, 279–293.
50. Rastelli, G.; Pacchioni, S.; Sirawaraporn, W.; Sirawaraporn, R.; Parenti, M. D.; Ferrari, A. M. Docking and database screening reveal new classes of *Plasmodium falciparum* dihydrofolate reductase inhibitors. *J. Med. Chem*. **2003**, 46, 2834–2845.
51. Chiang, R. A.; Sali, A.; Babbitt, P. C. Evolutionarily conserved substrate substructures for automated annotation of enzyme superfamilies. *PLoS Comput. Biol*. **2008**, 4, e1000142.
52. Bostrom, J.; Hogner, A.; Schmitt, S. Do structurally similar ligands bind in a similar fashion? *J. Med. Chem*. **2006**, 49, 6716–6725.
53. Jorgensen, W. L. Efficient drug lead discovery and optimization. *Acc. Chem. Res*. **2009**, 42, 724–733.

54. Lee, M. S.; Olson, M. A. Calculation of absolute protein–ligand binding affinity using path and endpoint approaches. *Biophys. J*. **2006**, 90, 864–877.
55. Bursulaya, B. D.; Totrov, M.; Abagyan, R.; Brooks, C. L., 3rd. Comparative study of several algorithms for flexible ligand docking. *J. Comput. Aided Mol. Des*. **2003**, 17, 755–763.
56. Cole, J. C.; Murray, C. W.; Nissink, J. W.; Taylor, R. D.; Taylor, R. Comparing protein–ligand docking programs is difficult. *Proteins* **2005**, 60, 325–332.
57. Cummings, M. D.; DesJarlais, R. L.; Gibbs, A. C.; Mohan, V.; Jaeger, E. P. Comparison of automated docking programs as virtual screening tools. *J. Med. Chem*. **2005**, 48, 962–976.
58. Perola, E.; Walters, W. P.; Charifson, P. S. A detailed comparison of current docking and scoring methods on systems of pharmaceutical relevance. *Proteins* **2004**, 56, 235–249.
59. Wang, R.; Lu, Y.; Wang, S. Comparative evaluation of 11 scoring functions for molecular docking. *J. Med. Chem*. **2003**, 46, 2287–2303.
60. Warren, G. L.; Andrews, C. W.; Capelli, A. M.; Clarke, B.; LaLonde, J.; Lambert, M. H.; Lindvall, M.; Nevins, N.; Semus, S. F.; Senger, S.; et al. A critical assessment of docking programs and scoring functions. *J. Med. Chem*. **2006**, 49, 5912–5931.
61. Wang, J.; Kang, X.; Kuntz, I. D.; Kollman, P. A. Hierarchical database screenings for HIV-1 reverse transcriptase using a pharmacophore model, rigid docking, solvation docking, and MM-PB/SA. *J. Med. Chem*. **2005**, 48, 2432–2444.
62. Friesner, R. A.; Banks, J. L.; Murphy, R. B.; Halgren, T. A.; Klicic, J. J.; Mainz, D. T.; Repasky, M. P.; Knoll, E. H.; Shelley, M.; Perry, J. K.; et al. Glide: a new approach for rapid, accurate docking and scoring. 1. Method and assessment of docking accuracy. *J. Med. Chem*. **2004**, 47, 1739–1749.
63. Friesner, R. A.; Murphy, R. B.; Repasky, M. P.; Frye, L. L.; Greenwood, J. R.; Halgren, T. A.; Sanschagrin, P. C.; Mainz, D. T. Extra precision glide: docking and scoring incorporating a model of hydrophobic enclosure for protein-ligand complexes. *J. Med. Chem*. **2006**, 49, 6177–6196.
64. Schnecke, V.; Kuhn, L. A. Virtual screening with solvation and ligand-induced complementarity. *Perspect. Drug Discov*., **2000**, 171–190.
65. Verdonk, M. L.; Chessari, G.; Cole, J. C.; Hartshorn, M. J.; Murray, C. W.; Nissink, J. W.; Taylor, R. D.; Taylor, R. Modeling water molecules in protein-ligand docking using GOLD. *J. Med. Chem*. **2005**, 48, 6504–6515.
66. Kalyanaraman, C.; Bernacki, K.; Jacobson, M. P. Virtual screening against highly charged active sites: identifying substrates of alpha-beta barrel enzymes. *Biochemistry* **2005**, 44, 2059–2071.
67. Huang, N.; Kalyanaraman, C.; Irwin, J. J.; Jacobson, M. P. Physics-based scoring of protein–ligand complexes: enrichment of known inhibitors in large-scale virtual screening. *J. Chem. Inf. Model*. **2006**, 46, 243–253.

68. Lyne, P. D.; Lamb, M. L.; Saeh, J. C. Accurate prediction of the relative potencies of members of a series of kinase inhibitors using molecular docking and MM-GBSA scoring. *J. Med. Chem.* **2006**, 49, 4805–4808.
69. Lee, M. R.; Sun, Y. Improving docking accuracy through molecular mechanics generalized Born optimization and scoring. *J. Chem. Theory Comput.* **2007**, 3, 1106–1119.
70. Perola, E. Minimizing false positives in kinase virtual screens. *Proteins* **2006**, 64, 422–435.
71. Graves, A. P.; Shivakumar, D. M.; Boyce, S. E.; Jacobson, M. P.; Case, D. A.; Shoichet, B. K. Rescoring docking hit lists for model cavity sites: predictions and experimental testing. *J. Mol. Biol.* **2008**, 377, 914–934.
72. Friesner, R. A.; Guallar, V. Ab initio quantum chemical and mixed quantum mechanics/molecular mechanics (QM/MM) methods for studying enzymatic catalysis. *Annu. Rev. Phys. Chem.* **2005**, 56, 389–427.
73. Senn, H. M.; Thiel, W. QM/MM studies of enzymes. *Curr. Opin. Chem. Biol.* **2007**, 11, 182–187.
74. Shurki, A.; Warshel, A. Structure/function correlations of proteins using MM, QM/MM, and related approaches: methods, concepts, pitfalls, and current progress. *Adv. Protein. Chem.* **2003**, 66, 249–313.
75. Brooks, B.; Karplus, M. Harmonic dynamics of proteins: normal modes and fluctuations in bovine pancreatic trypsin inhibitor. *Proc. Natl. Acad. Sci. USA* **1983**, 80, 6571–6575.
76. Levitt, M.; Sander, C.; Stern, P. S. Protein normal-mode dynamics: trypsin inhibitor, crambin, ribonuclease and lysozyme. *J. Mol. Biol.* **1985**, 181, 423–447.
77. Ma, J. Usefulness and limitations of normal mode analysis in modeling dynamics of biomolecular complexes. *Structure* **2005**, 13, 373–380.
78. de Groot, B.; van Aalten, D.; Scheek, R.; Amadei, A.; Vriend, G.; Berendsen, H. Prediction of protein conformational freedom from distance constraints. *Proteins Struc. Funct. Bioinf.* **1997**, 29, 240–251.
79. Seeliger, D.; Haas, J.; de Groot, B. L. Geometry-based sampling of conformational transitions in proteins. *Structure (London, England: 1993)* **2007**, 15, 1482–1492.
80. Sherman, W.; Day, T.; Jacobson, M. P.; Friesner, R. A.; Farid, R. Novel procedure for modeling ligand/receptor induced fit effects. *J. Med. Chem.* **2006**, 49, 534–553.
81. Wolfenden, R. Analog approaches to the structure of the transition state in enzyme reactions. *Acc. Chem. Res.* **1972**, 5, 10–18.
82. Wolfenden, R. Thermodynamic and extrathermodynamic requirements of enzyme catalysis. *Biophys. Chem.* **2003**, 105, 559–572.
83. Juhl, P. B.; Trodler, P.; Tyagi, S.; Pleiss, J. Modelling substrate specificity and enantioselectivity for lipases and esterases by substrate-imprinted docking. *BMC Struct. Biol.* **2009**, 9, 39.

84. Schramm, V. L. Enzymatic transition states and transition state analogues. *Curr. Opin. Struct. Biol.* **2005**, 15, 604–613.
85. Gutteridge, A.; Thornton, J. Conformational changes observed in enzyme crystal structures upon substrate binding. *J. Mol. Biol.* **2005**, 346, 21–28.
86. English, B. P.; Min, W.; van Oijen, A. M.; Lee, K. T.; Luo, G.; Sun, H.; Cherayil, B. J.; Kou, S. C.; Xie, X. S. Ever-fluctuating single enzyme molecules: Michaelis–Menten equation revisited. *Nat. Chem. Biol.* **2006**, 2, 87–94.
87. Warshel, A.; Florián, J. Computer simulations of enzyme catalysis: finding out what has been optimized by evolution. *Proc. Natl. Acad. Sci. USA* **1998**, 95, 5950–5955.
88. Nobeli, I., Favia, A. D.; Thornton, J. M. Protein promiscuity and its implications for biotechnology. *Nat. Biotechnol.* **2009**, 27, 157–167.
89. O'Brien, P. J.; Herschlag, D. Catalytic promiscuity and the evolution of new enzymatic activities. *Chem. Biol.* **1999**, 6, R91–R105.
90. Copley, S. D. Enzymes with extra talents: moonlighting functions and catalytic promiscuity. *Curr. Opin. Chem. Biol.* **2003**, 7, 265–272.
91. Jeffery, C. J. Moonlighting proteins. *Trends Biochem. Sci.* **1999**, 24, 8–11.
92. Khersonsky, O.; Roodveldt, C.; Tawfik, D. S. Enzyme promiscuity: evolutionary and mechanistic aspects. *Curr. Opin. Chem. Biol.* **2006**, 10, 498–508.
93. Hult, K.; Berglund, P. Enzyme promiscuity: mechanism and applications. *Trends Biotechnol.* **2007**, 25, 231–238.
94. Saghatelian, A.; Trauger, S. A.; Want, E. J.; Hawkins, E. G.; Siuzdak, G.; Cravatt, B. F. Assignment of endogenous substrates to enzymes by global metabolite profiling. *Biochemistry* **2004**, 43, 14332–14339.
95. Babbitt, P. C. Definitions of enzyme function for the structural genomics era. *Curr. Opin. Chem. Biol.* **2003**, 7, 230–237.
96. Furnham, N.; Garavelli, J. S.; Apweiler, R.; Thornton, J. M. Missing in action: enzyme functional annotations in biological databases. *Nat. Chem. Biol.* **2009**, 5, 521–525.
97. Ashburner, M.; Ball, C. A.; Blake, J. A.; Botstein, D.; Butler, H.; Cherry, J. M.; Davis, A. P.; Dolinski, K.; Dwight, S. S.; Eppig, J. T.; et al. Gene ontology: tool for the unification of biology. The Gene Ontology Consortium. *Nat. Genet.* **2000**, 25, 25–29.

CHAPTER 9

USING CHEMICAL STRUCTURE AS PROBES FOR BIOLOGICAL NETWORKS

FLORIAN NIGSCH
Chemical Biology Informatics, Quantitative Biology, Departmental and Molecular Pathways, Novartis Institutes for BioMedical Research, Cambridge, Massachusetts

1 INTRODUCTION

Since the beginning of this new century, there has been a trend in the life sciences to focus on integrated and systems approaches.[1,2] The precise meaning of *integrated* and *systems* often depends on the particular area of concern.[3] The mainstay of all of these approaches, however, is a shift in how biological organisms and external interventions in these are conceptualized. Much of this shift is due to the realization that the constituents of an organism such as a cell are not functioning in isolation.[4,5] Rather, they function together in interdependent cascades of chemical reactions that bring about the possibility for an organism to exert the functions necessary to sustain life. These cascades of reactions that take place in living cells are often referred to as *biological pathways*. Molecular biology has long been concerned with the elucidation of these pathways in healthy and diseased cells.[1] With advances in analytical techniques and improvements in the protocols of cellular molecular biology, the rate of new discoveries has increased. This wealth of data also results in the establishment of many cross-links between proteins and/or pathways that earlier were thought to be independent of each other. The resulting network of biological agents increasingly showcases the need to give up the notion of a single protein, enzyme, or even pathway in a cell as an isolated unit

Computational Approaches in Cheminformatics and Bioinformatics, First Edition.
Edited by Rajarshi Guha and Andreas Bender.

that carries out its task in a self-sufficient manner.[6] In this chapter we discuss what biological networks are, how they can be elucidated, what their properties are, and what they can be used for in drug discovery.

2 WHAT ARE BIOLOGICAL NETWORKS?

In the broadest sense, a network is a set of elements that are connected to each other with links. In the case of biological networks, these elements are molecules and proteins that are constituents of biological cells. The links between two elements correspond to the interactions that link them together.

The term *biological pathway* is commonly employed to refer to a medium-sized biological network that fulfills a well-defined function in a cell. Conceptually, it can be thought of as a series of biochemical reactions that transform an input molecule into a product molecule. Because each pathway is normally focused on a specific function, a cell needs a multitude of pathways to allow it to perform all the functions that it needs to survive. Such a collection of pathways therefore makes up the entire wiring of a cell, resulting in a highly complex biological network.

A practical example of a biological pathway that plays a major role in metabolism is fat catabolism (i.e., the breakdown of fat into smaller molecules and the generation of energy). This is one of the major pathways that converges toward the tricarboxylic acid (TCA) cycle that produces energy in the form of ATP. The triglycerides that make up fat are hydrolyzed into their constituent glycerol and fatty acids. By way of gluconeogenesis, a different pathway in the liver, glycerol can be converted into glucose. Fatty acids, on the other hand, can be broken down through a process known as *beta-oxidation*. This reaction yields acetylcoenzyme A, which can be used directly in the TCA cycle. A process known as oxidative phosphorylation is needed to extract the energy from the cycle in order for it to continue. This example shows briefly how a biological pathway carries out a specific function (digestion of fat to gain energy) through consecutive steps in a cascade of biochemical reactions that transform an input molecule (in this case, triglycerides) into a resulting output molecule (carbon dioxide). A main aspect of pathways is that often it is not only the input and output molecules that matter, but also any side effects that may occur. In the example above, this would be the reduction of a molecule of oxygen and the synthesis of two molecules of ATP with the resulting energy.

The fact that each pathway has been shaped through evolution to perform a specific function means that it has evolved into a subcellular processing unit that is indispensable. Therefore, if a pathway fails to fulfill its function, the cell will ultimately be put out of its normal state.

3 THE RELEVANCE OF BIOLOGICAL NETWORKS

As described in Section 2, a biological network that ultimately describes all processes that are taking place in a cell is made up of smaller units called biological pathways. Each of these units is required to keep the cell in a healthy state. Thus, a failure in any one pathway will result in an alteration of the state of the cell. Depending on the type of failure, the cell will either be able to perform an equivalent function via a different pathway, or when that is not possible, the cell will be imbalanced or even rendered nonfunctioning.[7] The result of such an imbalance or impairment of cellular function is often the cause of a disease.[8]

An example for such an imbalance would be the overexpression of receptor protein–tyrosine kinases.[9] Their constitutive overexpression leads to an accummulation of these proteins on the cell surface, which in turn leads to constitutive kinase activation through cross-phosphorylation of two kinases in proximity. Prominent examples for diseases caused by this mechanism are breast and lung carcinomas, linked to Neu/ErbB2 (erythroblastic leukemia viral oncogene homolog 2) and epidermal growth factor receptor (EGFR).[10]

Another example of a well-studied pathway that is implicated in different diseases is the Wnt/beta-catenin signaling pathway. It regulates adult tissue homeostasis and regeneration, as well as cell fate and behavior during embryogenesis.[11] Correct levels of stimulation of this pathway are very important, and deviations from this baseline in either direction have been linked to diseases: inappropriate activation leads to colorectal cancer and melanoma,[12] whereas a lack of stimulation may contribute to Alzheimer's disease and osteoporosis.[13,14]

As can be seen from this example, a single pathway may be implicated in many seemingly unrelated processes in healthy, as well as diseased, cells. Furthermore, depending on the state of the pathway, activation or repression, its effects farther downstream may also change. This has implications in the case of a divergent pathway when one downstream branch should be inhibited for therapeutic intervention, whereas the other branch, if inhibited, may adversely affect the therapeutic outcome. An example of this would be inhibition of TGF-beta receptor (TGFR). Downstream of this receptor are SMAD2 and SMAD3, both implicated in cell growth, as well as the mitogen-activated protein kinase (MAPK) cascade.[15,16]

The preceding paragraphs demonstrate the importance of connected biological networks in terms of the implications they have for both homeostasis and therapeutic interventions.

4 BIOLOGICAL NETWORK ELUCIDATION

The elucidation of biological networks can be accomplished in many different ways. There is, however, no single approach that is superior to all others. Rather, every approach has strengths and weaknesses. One of the first methods used to study protein–protein interactions was the yeast two-hybrid technique (yeast two-hybrid screening, Y2H).[17] The principle behind this method is the activation of a transcription factor that is split into two parts, the binding domain and the activating domain. Each of these can be attached to a separate protein. When the binding domain binds to its DNA target, and subsequent binding of its attached protein to the protein with the activating domain occurs, the reporter gene downstream of the transcription factor will be expressed. Thus, by means of expression of the reporter gene it is possible to identify the interaction of two proteins.[18]

The yeast two-hybrid assay has its shortcomings, one of the most important being its high rate of false positives (and false negatives). Other techniques that can be used to validate or invalidate the results of the yeast two-hybrid system include coaffinity purification in combination with mass spectrometry.[19,20] Both of these techniques, however, were shown to have a similar rate of false positive discovery.[21] Therefore, they can be used to derive consensus results (i.e., keep those interactions that have been identified independently in different assays). Verification of interactions with methods of higher confidence, such as coimmuniprecipitation assays, should always be considered.[22] Because of the increased experimental complexity and effort required in the setup and execution of such assays, these techniques are often only applicable on a smaller scale than the less accurate techniques.

Apart from the experimental techniques mentioned in the preceding paragraph, another method that is often used to establish relationships among proteins is text mining.[23,24] In this approach the abstracts of thousands of publications referring to, for example, a particular disease are obtained from PubMed.[25] In a subsequent processing step, all co-occurrences of proteins are identified. With these data in hand, it is possible to construct a network of all the proteins that are—presumably—linked in the context of the specific disease. This same approach is obviously also applicable for other contexts, such as specific cell signaling events (e.g., immunostimulant activity of certain receptors).[26] This method is a useful approach to construct networks of particular interest which may not be probed experimentally. Furthermore, since this approach gathers many independent observations of molecular interactions, it is possible to identify interactions that gain in significance only in the context of large volumes of information. Such interactions may be disseminated over the scientific literature and, because of that, never put into a common context that would allow this interaction to be recognized as important in, for example, a specific cellular context, be it disease or other.[27]

The same reasons that confer this strength in identifying unanticipated relationships can also be a drawback: The co-occurence of two terms is not in itself a reason for these two proteins to be related in a biologically meaningful way. It could well be that two proteins are mentioned sufficiently often together to be identified as interacting proteins. Yet they may be expressed in different compartments of the cell at different, nonoverlapping times and therefore not be related in a biologically meaningful way.

5 BIOLOGICAL NETWORK PROBING

In the previous paragraphs we talked about general aspects of network elucidation. In the next paragraphs we focus on more specific approaches, in particular to what extent small biological molecules can be put to use as probes for biological networks.

Small-Molecule Probes

Many endogenous biological functions and processes rely on small molecules as carriers of information. For example, the family of molecules derived from lanosterol, the precursor of the steroids, is responsible for the regulation of a number of important processes.[28] These include developmental events such as sex differentiation and reproduction (sex steroids), regulation of metabolism and immune function (corticosteroids), and muscle and bone growth (anabolic steroids).

In contrast to proteins as carriers of biological information, small molecules have to rely on many fewer, yet more specific interactions with their intended target. Molecular recognition events between two proteins typically occur at interaction surfaces which are orders of magnitude larger than those found in interactions between a small molecule (endogenous or xenobiotic) and its receptor(s).[29] This has important implications for the design and use of small (organic) molecules as probes for biological networks.

As an illustration, consider a protein–protein interaction as the question of interest. There may be dozens of amino acid residues partaking in the interaction, and the total surface that is involved in the binding of the two proteins may well span several hundred square angstroms. The intermolecular forces generated at this interacting surface are far beyond what is achievable by a small molecule–protein interaction.

Two immediate consequences of this observation are the fact that (1) small-molecule probes for any biological function need to be highly specific to enable targeting of the desired protein; and (2) the interrogation or disruption of protein–protein interactions is a very difficult task, due to the size differences of the binding partners and the resulting differences in forces that can be generated. It is therefore of prime importance to have small molecular probes of high affinity to their targets.[30]

One category of such high-affinity probes to be cited is the class of molecules manufactured by evolution in the form of natural products.[31,32] Natural products have been shaped by evolutionary design to exert highly specific functions in the systems in which they operate. These highly specific functions translate into a very high affinity for a certain receptor and not for others.

An example of such a compound is epibatidine. In the 1970s a compound was extracted from the frog species *Epipedobates tricolor*, which had very potent analgesic properties.[33] Due to a lack of material, however, the structure of this compound was not determined until the introduction of more sensitive NMR instruments in the early 1990s. The structure was then determined to be (1*R*, 2*R*, 4*S*-exo-2-(6-chloro-3-pyridyl)-7 -azabicyclo[2.2.1]heptane). In addition to its analgesic potential (approximately 200 times as active as morphine), this compound was also believed not to act at the opioid receptors (Figure 1).[34] It was later found that epibatidine binds to the nicotinic acetylcholine receptors (nAChRs). Moreover, it does so with very high specificity at one of the subtypes of these receptors. The neuronal nicotinic acetylcholine receptors come in several subtypes, composed of alpha and beta subunits.[35] Epibatidine was shown to have picomolar activity at several of these subtypes: for example, 10.3 and 13.6 pM at the $\alpha2\beta2$ and $\alpha3\beta2$ types, respectively (Table 1).[36]

The use of small molecules to alter the function of proteins in cells (often referred to as chemical genetics) has several advantages over traditional genetic approaches.[37] In the latter, genes can be silenced (knocked-out), which prevents the cell from producing the corresponding protein product. The necessary result of such a knock-out experiment is rather extreme: There is none of the targeted protein in the system under scrutiny.

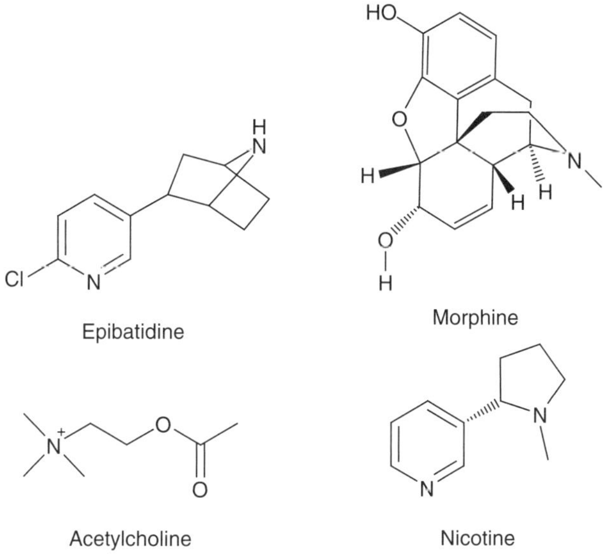

FIGURE 1 Epibatidine and related compounds.

TABLE 1 Picomolar Activities of Epibatidine at Nicotinic Acetylcholinergic Receptor Subtypes (nAChR)[a]

α2β2	α3β2	α4β2
10.3	13.6	30.0
α2β4	α3β4	α4β4
86.8	303	84.7

[a]Various combinations of the alpha and beta subunits of the receptor are shown.

In other words, traditional genetics experiments are binary in nature: Either a protein is present (in the wild-type system) or absent (in the genetically modified system). With chemical genetics, however, this is not the case because the degree of inhibition can be controlled both in magnitude and temporal fashion. More specifically, individual functions of multifunctional proteins can be selectively modulated with small molecules. This would never be possible with the traditional genetic silencing of the entire protein.

In forward chemical genetics a library of small molecules is tested against a particular cell line, and any phenotype-modifying effects of the compounds are registered.[38,39] Compounds that elicit the desired phenotypic response may then be selected for follow-up experiments. In reverse chemical genetics, the constituents of a small molecule library are anchored on solid support media (e.g., polyacetate resin or slide) to yield the equivalent of a DNA microarray chip where the probes are small molecules. The labeled protein target of interest is then applied to the entire resin. Analysis of the molecule chip then allows identification of those molecules that bind to the protein of interest. These molecules can then be subjected to further detailed in vitro validation in individual screens to verify their activity against the protein target.[20]

An early analysis of this kind was carried out by Schreiber and colleagues. They used diversity-oriented synthesis (DOS) to construct a small-molecule library of 3780 structurally complex compounds sharing a 1,3-dioxane scaffold.[40] This library was immobilized on a microarray chip and screened in a reverse chemical genetic screen against Ure2p. This protein is a nutrient-responsive yeast signaling protein and acts as a repressor of genes involved in nitrogen metabolism. Moreover, it can also form prions and is part of the signaling cascade downstream of the Tor proteins. The reverse chemical genetic screen identified eight protein-binding molecules that were resynthesized and tested in a secondary cell-based assay. Of the eight compounds, only one (named uretupamine A) showed activity in this assay. With the use of surface plasmon resonance (SPR) studies, this compound was then identified to have a K_D value of 18.1 μM for Ure2p. In an effort to yield a more potent inhibitor, a close analog was synthesized that had a K_D value of 7.5 μM. One advantage of diversity-oriented synthesis is that structural analogs are readily obtained, as the synthetic routes used in the first place are already established.

Instead of using only one small molecule to elicit a certain response in a cell, screens that use two or more small molecules at the same time have also been carried out. The concurrent use of two compounds is commonly employed in clinical settings when drug combinations are used for the treatment of certain diseases.[41] In drug discovery efforts, however, the focus is mostly toward single chemical entities.

One early example of systematic screening of combinations of bioactive substances was demonstrated by Haggarty et al. in 2003.[42] In a setting that they called chemical genomic profiling, they tested 24 bioactive compounds in pairwise combinations against a wild-type strain and nine additional isogenic strains of yeast (*Saccharomyces cerevisiae*), each carrying a single gene deletion of a different component of the spindle assembly/cell polarity network.[43] The phenotypic readout of the treatment of each strain with one of the possible combinations was optical density, as it is directly related to cell growth in medium. Every combination that resulted in change of growth—either no growth or reduced growth—was given a positive score of 1. No change in growth was assigned a neutral score of 0. This allowed the construction of an adjacency matrix of the effects of all combinations of the 24 molecules on each strain. Subsequent analysis of these adjacency matrices allowed the identification of highly connected molecules (identified by optical inspection of a force-directed layout), and also the derivation of molecule-specific fingerprints for each molecule in each strain.

For each strain-specific network, Haggarty et al. calculated graph-based descriptors akin to those used in QSAR analysis. These descriptors were then used to cluster the networks based on their similarity with respect to these descriptors. It was found that this clustering was biologically meaningful, as the cin8Δ and kip1Δ deletion strains ended up being nearer to each other than to any other strain. The two genes CIP8 and KIP1 both play a genetically redundant role in the function and organization of the mitotic spindle during cell division.[43]

Similarly, the information contained in the adjacency matrix for each molecule was used to cluster the molecules. This allows a clustering of molecules based on the similarities in their pattern of biological activity. The molecules closest to each other using this approach were the same ones that have been identified in the earlier analysis as being the most highly connected. Overall, this method was shown to be useful in the analysis of the diversity of small molecules by systematically varying the genetic networks against which they are tested.

A more recent study that also focused on the effects of pairwise combinations was presented by Yeh et al.[44] They selected a set of 21 known antibiotics and tested them in pairwise fashion for inhibition of growth of *Escherichia coli* bacteria. Instead of focusing on the interactions of specific pairs of drugs, the prime interest of this study was the analysis of the full drug interaction network. The 21 drugs were chosen to reflect different mechanisms of action (e.g., cell wall synthesis, nucleic acid synthesis, protein synthesis, and folic acid synthesis). Contrary to a clinical setting where antibiotic drugs are typically used at high doses, they focused on sublethal concentrations to be closer to the physiological state of the wild-type bacteria.[45] Pairwise interaction experiments were then performed with concentrations of drugs that induced between 50 and 90% growth inhibition. All drugs were then assayed individually and in combination to identify potential synergistic or antagonistic effects arising from coadministration of the drugs. Each drug pair was assigned one of three effects: (1) synergistic (synthetic lethal interaction): no growth when exposed to both drugs; (2) antagonistic buffering: one drug completely masks the effect of the other; and (3) antagonistic suppression: the combination yields a higher growth rate than that of one drug alone.

Having assigned a type to each combination of drugs, they then used a clustering algorithm (PRISM) to divide the set of all drugs into groups

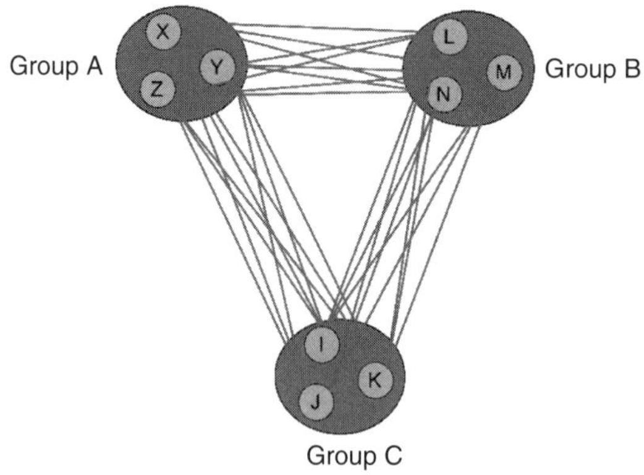

FIGURE 2 Monochromatically interacting groups of molecules. Two groups are interacting monochromatically if all molecules of a group interact in the same fashion when combined with any of the molecules of the other group.

of drugs that interact only "monochromatically" with each other.[46] Two groups of drugs, A and B, are defined to interact monochromatically when all drugs of groups A have the same type of effect when combined with drugs from group B. For example, if all combinations of drugs of groups A and B would have a synergistic effect, the interaction between groups A and B would be monochromatic (Figure 2).

It was found that the groups identified reflected very closely the putative mechanism of action of the drugs contained within a group. Examples of such clusters were sulfamonomethoxine and trimethoprim (folic acid synthesis inhibitors); piperacillin, ampicillin, cefoxitin (cell wall interference); and nalidixic acid, lomefloxacin, ciprofloxacin (DNA gyrase inhibitors). Moreover, the clustering obtained using the PRISM algorithm revealed that the resulting classification was more reflective of the cellular function rather than the structure of the compounds. For example, one of the drugs classified among the other 30S drugs is structurally an aminoglycoside. Apart from clustering molecules according to their biological function, such network approaches can also be used to uncover new functionalities of molecules. One of the drugs assayed by Yeh et al. did not exhibit monochromatic behavior: namely, the synthetic antibiotic nitrofurantoin. This drug is known to act on all of DNA synthesis, protein synthesis, and cell wall synthesis. The fact that this drug cannot be clustered into the network hints at the potential of this method to identify drugs that have multiple modes of action. Another way to exploit the fact that certain drugs cannot be attributed monochromatically to existing clusters consists of screening for novel modes of action. If a particular molecule cannot be attributed to any of the already existing clusters, this must be due to it having a different mechanism of action. In the study of Yeh et al., bleomycin, a drug with a less well characterized mechanism of action, was included. Bleomycin is an anticancer drug that also inhibits the growth of *E. coli*. The mechanism by which it does so is assumed to be related to oxidation and degradation of DNA. In the drug interaction network derived using the PRISM algorithm, bleomycin did indeed end up in a cluster of its own because of a mechanism of action that was distinct from all the other groups of drugs.

One of the largest efforts undertaken to date to profile comprehensively the responses of a large number of drugs against different cell lines has been accomplished by the Connectivity Map.[47] This work has been carried out at the Broad Institute, where Lamb et al. screened a diverse set of 164 small molecule perturbagens, including U.S. Food and Drug Administration–approved drugs and nondrug tool compounds. For each compound, a complete expression profile was obtained by analyzing the cellular response to treatment via microarray chips. The data obtained consist of full mRNA expression profiles of the 164 compounds across different cell lines, as well as control experiments.

The selection of compounds was done so as to include compounds that have similar modes of action, are indicated for treatment of the same disease (e.g., antidiabetics), act proximal or distal of gene expression (estrogen receptor modulators and immunomodulators, respectively), compounds whose targets are not expressed in all cell types (e.g., COX2 inhibitors), as well as compounds that manifest their effects only after chronic in vivo exposure (e.g., antipsychotics). Most of the compounds were screened in the breast cancer epithelial cell line MCF7. In addition, some of the compounds were also screened in the prostate cancer epithelial cell line PC3 and the nonepithelial lines HL60 (leukemia) and SKMEL5 (melanoma).

The data contained in the Connectivity Map can be queried through the use of gene set enrichment analysis (GSEA), which matches a supplied query signature to the data.[48] GSEA has several advantages over other methods, such as hierarchical clustering, which is often employed in the context of mRNA expression profiles. In short, the main advantages of GSEA are (1) more sensitivity to the treatment than to batch effects or effects relating to the cell line used in experiments; (2) independence of the platform with which the expression profiles were generated—this allows querying of the Connectivity Map with gene expression profiles that have been obtained on any platform; and (3) GSEA can detect multiple components of the cellular response to a perturbagen.

Lamb et al. were able to confirm relations between small molecules using the Connectivity Map. To validate their approach they first tried to recover compounds from their data with signatures of these compounds obtained from published gene signatures. They used a gene signature comprising 13 differentially expressed genes that resulted from the treatment of bladder or breast carcinoma cell lines with three different histone deacetylase inhibitors HDACs (varinostat, trichostatin A, MS-27-275).[49] The 13-gene signature was derived in different cell lines than those in which the Connectivity Map data was obtained, yet it was successful in retrieving varinostat and trichostatin A as the two top-scoring compounds from the Connectivity Map. Moreover, two structurally distinct compounds were also highly correlated to the HDAC inhibitory profile, suggesting the use of the Connectivity Map to provide mechanism of action for structurally diverse compounds.

In the case of estrogen receptor modulators, the authors showed that both positively as well as negatively correlated compounds were retrieved. When using a query derived from 17β-estradiol-treated MCF7 breast cancer cell lines, highly positively scoring results from the Connectivity Map were 17β-estradiol itself along with the phytoestrogen genistein and 17α-estradiol. More important, though, among the compounds with highly negatively correlated profiles were the antiestrogenic drugs fulvestrant, tamoxifen, and raloxifene.

The authors further went on to use the Connectivity Map to elucidate the mechanisms of action of a compound, gedunin, that was found active in a separate screen to abrogate androgen receptor activity.[50] A query signature was obtained from treated LNCaP prostate cancer cells and used to obtain matching profiles from the Connectivity Map. Several HSP90 inhibitors were found as top-scoring results, suggesting that gedunin might be active in the HSP90 pathway. In a separate confirmation experiment, gedunin was verified to abrogate androgen receptor expression as well as HSP90-related protein expression in LNCaP and BA/F3 cell lines. This showed that the Connectivity Map is indeed useful to provide testable hypotheses with respect to the modes of action of compounds.

Lamb et al. also used disease-specific gene signatures to query the Connectivity Map. When they used an expression signature related to an obese state, they were able to retrieve peroxisome proliferator-activated receptor gamma (PPARγ) agonists. The retrieved thiazolidinediones (troglitazone and rosiglitazone) are used as antidiabetics, but one of their side effects in clinical use is the promotion of weight gain.[47] This example further validated the approach taken by Lamb et al. for use of the Connectivity Map in a disease-related context.

RNAi

There is a category of molecules that is somewhat between the large molecules such as proteins and the traditionally druglike molecules with molecular weight around 500. The molecules in this class are short interfering RNA molecules. These are highly specific molecular probes and therefore deserve brief mention here.

The naturally occurring process of RNA interference was discovered in the late 1990s when it was observed that the presence of double-stranded RNA (dsRNA) in a cell results in the degradation of the mRNA matching that sequence.[51] This discovery opened new doors for reverse genetics experiments whereby specific genes could be silenced through the introduction of dsRNA corresponding to the protein to be eliminated. The drawback of the introduction of dsRNA into mammalian cellular systems, however, is the induction of an antiviral interferon response that leads to subsequent cell death.[52] A breakthrough to alleviate this shortcoming was the discovery that a first step in RNA interference was the cleavage of dsRNA into smaller fragments consisting of 21 to 23 base pairs; accordingly, these fragments were named *small interfering RNAs* (siRNAs).[53] The siRNAs can be introduced into mammalian cells without activating an antiviral response.

siRNA technologies have been developed to handle large-scale high-throughput screens in which libraries of tens of thousands of siRNAs can be screened. Even though siRNAs are very specific, such large-scale screens are also not free of the problems of false-negative and false-positive results. There are certain limiting factors inherent in siRNA screening, some better understood than others. For example, the level of silencing that can be obtained depends significantly on the targeted mRNA sequence.[54] Moreover, there is also variability that cannot be explained by the sequence composition but which may, rather, depend on one or more of secondary structure, mRNA turnover, mRNA abundance, or other factors. One of the techniques employed to date to increase efficiency and accuracy in siRNA experiments is the use of several siRNAs of different sequence that target the same mRNA.

Screening with siRNA has proven to be a valuable tool to dissect pathways and continues to do so. Notable examples include cancer-associated screens of the Hedgehog (Hh) and Wingless (Wg) pathways of *Drosophila*,[55] as well as a screen targeting 30 components of the human PI3K pathway.[56] The latter identified the two known regulators of Akt phosphorylation, PTEN and PDK1. Berns et al. screened a library of approximately 24,000 retroviral vector-encoded short hairpin

RNAs (shRNAs) targeted against 7914 human genes to identify new components of the p53 pathway.[57]

Using Biological Networks in Drug Discovery

The concept of networked biological systems also finds progressively wider application in drug discovery settings. Some of the examples mentioned above, such as the Connectivity Map and the studies performed by Schreiber et al. and Yeh et al. clearly demonstrate the potential of such approaches.

Another type of experiment that probes biological networks are cell-based phenotypic screens. When a cell is subjected to compound treatment, the response can only be systemic in nature, and this manifests itself through the phenotype of the treated cells.[38] This experimental outcome can vary significantly depending on the level of detail with which phenotypes are defined. If, for example, the phenotype is growth inhibition of bacteria, it is a rather high-level endpoint. At the other end of the scale one finds high-content phenotypic screens, which may involve highly sensitive imaging techniques that can capture deformations of nuclei, changes in cellular motility, and so on. Because of the data generation rate, which is much higher with respect to the lower-grain endpoints (compare the determination of optical density for bacterial growth with multiple high-resolution single-cell images), such experiments are still somewhat limited in throughput to between hundreds and thousands of compounds.[58]

Clear advantages for cell-based phenotypic screens are that they inherently address questions of compound solubility, cell membrane permeability, and cytotoxicity. A drawback, however, is the significant effort that goes into the determination of the actual targets that cause a desired phenotype of interest. Knowledge about biological networks can help to limit the search space upfront by focusing on pathways that are likely to lead to the phenotype observed. It is, however, also possible to take an opposite approach in which screening libraries are designed that are tailored to a specific pathway.

Knowledge about the connectivities of drugs in biological networks also allows for the identification or rationalization of potential side effects of drugs. The concept that side effects of drugs can actually be exploited to good use by repurposing them as drugs for the observed off-target is not new. Wermuth published a report in 2006 describing the selective optimization of side activities.[59] With the availability of large-scale chemogenomics and pathways databases, analyses that can be carried out today look at that challenge from a different point of view. The information contained in large databases of compound activity information can be used to construct drug interaction networks, drug–target networks, or drug–disease networks, which can all be used in various ways for different purposes.

Scheiber's group used pathways to cxplain off-target effects of compounds by analyzing the predicted targets of compounds with known adverse drug reactions (ADRs) in the context of the pathways of these targets.[60] Yang et al. used a docking approach and a gene co-citation network to explain serious adverse drug reactions.[61] Analyses of drug-interaction networks have also been used to explain and predict off-target effects of drugs.[62,63]

Focusing less on small organic molecules and more on genes and proteins, several groups have constructed disease-specific networks. Karni et al. integrated protein–protein interaction data with gene expression data obtained under certain disease conditions, and they used their approach to predict disease-causing genes.[64] Dezso et al. took a different approach, in which they analyzed genes with respect to their topological properties in disease-related contexts. They were able to confirm several validated disease-implicated genes for psoriasis, as well as to propose new targets.[8] Qu et al. built a disease–drug correlation ontology and analyzed the resulting RDF graph.[65] Using systemic lupus erythematosus (SLE) as an example, they were able to verify the validity of their model, which identified tamoxifen as the most likely drug to treat SLE—tamoxifen is indeed a current clinical candidate for the treatment of SLE.

An area of intense research that is intrinsically linked to knowledge about the networked design of cells is the search for multicomponent therapeutics, or combination drug therapies.[41] There are many examples of demonstrated higher clinical efficiency resulting from the combination and co-administration of two drugs. One particular case is the combination of gefitinib, an EGFR tyrosine kinase inhibitor, and taxane, a microtubule destabilizing agent. This combination was shown to have a highly synergistic effect in breast cancer cells.[41]

There are several mechanisms through which drug combinations can achieve increased efficiency,[41] including (1) modulating different targets in the same pathway, (2) modulating different targets of related pathways, (3) modulating different targets of cross-talking pathways, (4) modulating different targets in the same pathway that cross-talk by means of another pathway, and (5) modulation of different sites of the same target.

The use of combination drug treatments is warranted by the nature of the wiring of cellular systems. Through evolutionary design, there is robustness and redundancy built into many of the regulatory and signaling pathways found in higher organisms.[2,66] As a consequence, it is often beneficial to modulate more than one disease-implicated protein target.

Apart from co-modulation of protein targets, another use for drug combinations is to use a second drug to abrogate adverse effects related to the first one. This case is encountered in the combination of cisplatin and procainamide, where the latter is used to reduce cisplatin-induced hepatotoxicity.[41]

Combinations of molecules can also be used to explore connectivities of biological systems. Lehár et al. showed how the relationships between targets in a pathway can be inferred from the responses elicited by application of combinations of compounds.[67] They argue that such approaches can be used to complement genetic studies as well as probe disease models that are not amenable to traditional mutagenesis investigations.

Overall, there are many tangible reasons for the study of biological networks in the context of multiple small-molecule perturbations. These investigations can be seen from two angles: (1) combinations of molecules can be used to probe and elucidate biological networks, and (2) biological networks can be used to infer combinations of nodes that might have the most effect if co-modulated.

6 CONCLUSION

This discussion introduced some notions about biological networks and how they can be probed and modulated. This modulation can be achieved with small molecules, including drugs and/or tool compounds as well as siRNAs. Other possibilities for specific modulation of biological networks such as antibodies have not been discussed, as the focus here is on small molecules. Several examples of applications of specific techniques were presented in a wider context to illustrate the impact that such methods can have on both the elucidation of biological networks and on rationally designed, targeted modulation of these. An area of particular interest that is bound to accrue more research resources is the use of combinations of perturbations to biological networks. Especially in clinical settings, such approaches can have important implications for the rational modulation of diseased cells.

REFERENCES

1. Schadt, E. E.; Friend, S. H.; Shaywitz, D. A. A network view of disease and compound screening. *Nat. Rev. Drug Discov*. **2009**, 8, 286–295.

2. Hellerstein, M. K. A critique of the molecular target-based drug discovery paradigm based on principles of metabolic control: advantages of pathway-based discovery. *Metab. Eng*. **2008**, 10, 1–9.
3. Bugrim, A.; Nikolskaya, T.; Nikolsky, Y. Early prediction of drug metabolism and toxicity: systems biology approach and modeling. *Drug Discov. Today* **2004**, 9, 127–135.
4. Blow, N. Systems biology: untangling the protein web. *Nature* **2009**, 460, 415–418.
5. Oprea, T. I.; Tropsha, A.; Faulon, J.; Rintoul, M. D. Systems chemical biology. *Nat. Chem. Biol*. **2007**, 3, 447–450.
6. Li, Y.; Agarwal, P. A pathway-based view of human diseases and disease relationships. *PLoS ONE* **2009**, 4, e4346.
7. Wu, Y.; Zhang, X.; Yu, J.; Ouyang, Q. Identification of a topological characteristic responsible for the biological robustness of regulatory networks. *PLoS Comput. Biol*. **2009**, 5, e1000442.
8. Dezso, Z.; Nikolsky, Y.; Nikolskaya, T.; Miller, J.; Cherba, D.; Webb, C.; Bugrim, A. Identifying disease-specific genes based on their topological significance in protein networks. *BMC Syst. Biol*. **2009**, 3, 36.
9. Blume-Jensen, P.; Hunter, T. Oncogenic kinase signalling. *Nature* **2001**, 411, 355–365.
10. Harari, D.; Yarden, Y. Molecular mechanisms underlying ErbB2/HER2 action in breast cancer. *Oncogene* **2000**, 19, 6102–6114.
11. Zhang, Q.; Major, M. B.; Takanashi, S.; Camp, N. D.; Nishiya, N.; Peters, E. C.; Ginsberg, M. H.; Jian, X.; Randazzo, P. A.; Schultz, P. G.; Moon, R. T.; Ding, S. Small-molecule synergist of the Wnt/beta-catenin signaling pathway. *Proc. Natl. Acad. Sci. USA* **2007**, 104, 7444–7448.
12. Goel, A.; Boland, C. Recent insights into the pathogenesis of colorectal cancer. *Curr. Opin. Gastroenterol*. **2010**, 26, 47–52.
13. Boonen, R. A. C. M.; van Tijn, P.; Zivkovic, D. Wnt signaling in Alzheimer's disease: up or down, that is the question. *Ageing Res. Rev*. **2009**, 8, 71–82.
14. MacDonald, B. T.; Tamai, K.; He, X. Wnt/beta-catenin signaling: components, mechanisms, and diseases. *Dev. Cell* **2009**, 17, 9–26.
15. Law, A. K. T.; Gupta, D.; Levy, S.; Wallace, D. C.; McKeon, R. J.; Buck, C. R. TGF-beta1 induction of the adenine nucleotide translocator 1 in astrocytes occurs through Smads and Sp1 transcription factors. *BMC Neurosci*. **2004**, 5, 1.
16. Keith, C. T.; Borisy, A. A.; Stockwell, B. R. Multicomponent therapeutics for networked systems. *Nat. Rev. Drug Discov*. **2005**, 4, 71–78.
17. Lentze, N.; Auerbach, D. The yeast two-hybrid system and its role in drug discovery. *Expert Opin. Ther. Targets* **2008**, 12, 505–515.
18. Parrish, J. R.; Gulyas, K. D.; Finley, R. L. Yeast two-hybrid contributions to interactome mapping. *Curr. Opin. Biotechnol*. **2006**, 17, 387–393.

19. Calleri, E.; Temporini, C.; Caccialanza, G.; Massolini, G. Target-based drug discovery: the emerging success of frontal affinity chromatography coupled to mass spectrometry. *ChemMedChem* **2009**, 4, 905–916.

20. Rix, U.; Superti-Furga, G. Target profiling of small molecules by chemical proteomics. *Nat. Chem. Biol*. **2009**, 5, 616–624.

21. Huang, H.; Bader, J. S. Precision and recall estimates for two-hybrid screens. *Bioinformatics* **2009**, 25, 372–378.

22. Kay, B. K.; Williamson, M. P.; Sudol, M. The importance of being proline: the interaction of proline-rich motifs in signaling proteins with their cognate domains. *FASEB J*. **2000**, 14, 231–241.

23. Huang, M.; Ding, S.; Wang, H.; Zhu, X. Mining physical protein–protein interactions from the literature. *Genome Biol*. **2008**, 9(Suppl. 2), S12.

24. Bandy, J.; Milward, D.; McQuay, S. Mining protein–protein interactions from published literature using Linguamatics I2E. *Methods Mol. Biol*. **2009**, 563, 3–13.

25. Li, J.; Zhu, X.; Chen, J. Y. Building disease-specific drug–protein connectivity maps from molecular interaction networks and PubMed abstracts. *PLoS Comput. Biol*. **2009**, 5, e1000450.

26. Raychaudhuri, S.; Thomson, B. P.; Remmers, E. F.; Eyre, S.; Hinks, A.; Guiducci, C.; Catanese, J. J.; Xie, G.; Stahl, E. A.; Chen, R.; et al. Genetic variants at CD28, PRDM1 and CD2/CD58 are associated with rheumatoid arthritis risk. *Nat. Genet*. **2009**, 41, 1313–1318.

27. Krallinger, M.; Erhardt, R. A.; Valencia, A. Text-mining approaches in molecular biology and biomedicine. *Drug Discov. Today* **2005**, 10, 439–445.

28. Abe, I. Enzymatic synthesis of cyclic triterpenes. *Nat. Prod. Rep*. **2007**, 24, 1311–1331.

29. Janin, J.; Bahadur, R. P.; Chakrabarti, P. Protein–protein interaction and quaternary structure. *Q. Rev. Biophys*. **2008**, 41, 133–180.

30. Li, X.; Barasoain, I.; Matesanz, R.; Díaz, J. F.; Fang, W. Synthesis and biological activities of high affinity taxane-based fluorescent probes. *Bioorg. Med. Chem. Lett*. **2009**, 19, 751–754.

31. Breinbauer, R.; Manger, M.; Scheck, M.; Waldmann, H. Natural product guided compound library development. *Curr. Med. Chem*. **2002**, 9, 2129–2145.

32. Hung, D. T.; Jamison, T. F.; Schreiber, S. L. Understanding and controlling the cell cycle with natural products. *Chem. Biol*. **1996**, 3, 623–639.

33. Daly, J. W.; Garraffo, H. M.; Spande, T. F.; Decker, M. W.; Sullivan, J. P.; Williams, M. Alkaloids from frog skin: the discovery of epibatidine and the potential for developing novel non-opioid analgesics. *Nat. Prod. Rep*. **2000**, 17, 131–135.

34. Badio, B.; Daly, J. W. Epibatidine, a potent analgetic and nicotinic agonist. *Mol. Pharmacol*. **1994**, 45, 563–569.

35. Brioni, J. D.; Decker, M. W.; Sullivan, J. P.; Arneric, S. P. The pharmacology of (−)-nicotine and novel cholinergic channel modulators. *Adv. Pharmacol*. **1997**, 37, 153–214.

36. Parker, M. J.; Beck, A.; Luetje, C. W. Neuronal nicotinic receptor beta2 and bcta4 subunits confer large differences in agonist binding affinity. *Mol. Pharmacol*. **1998**, 54, 1132–1139.

37. Tan, D. S. Sweet surrender to chemical genetics. *Nat. Biotechnol*. **2002**, 20, 561–563.

38. Feng, Y.; Mitchison, T. J.; Bender, A.; Young, D. W.; Tallarico, J. A. Multiparameter phenotypic profiling: using cellular effects to characterize small-molecule compounds. *Nat. Rev. Drug Discov*. **2009**, 8, 567–578.

39. Shim, J. S.; Kwon, H. J. Chemical genetics for therapeutic target mining. *Expert Opin. Ther. Targets* **2004**, 8, 653–661.

40. Kuruvilla, F. G.; Shamji, A. F.; Sternson, S. M.; Hergenrother, P. J.; Schreiber, S. L. Dissecting glucose signalling with diversity-oriented synthesis and small-molecule microarrays. *Nature* **2002**, 416, 653–657.

41. Jia, J.; Zhu, F.; Ma, X.; Cao, Z.; Cao, Z. W.; Li, Y.; Li, Y. X.; Chen, Y. Z. Mechanisms of drug combinations: interaction and network perspectives. *Nat. Rev. Drug Discov*. **2009**, 8, 111–128.

42. Haggarty, S. J.; Clemons, P. A.; Schreiber, S. L. Chemical genomic profiling of biological networks using graph theory and combinations of small molecule perturbations. *J. Am. Chem. Soc*. **2003**, 125, 10543–10545.

43. Segal, M.; Bloom, K. Control of spindle polarity and orientation in *Saccharomyces cerevisiae*. *Trends Cell Biol*. **2001**, 11, 160–166.

44. Yeh, P. J.; Hegreness, M. J.; Aiden, A. P.; Kishony, R. Drug interactions and the evolution of antibiotic resistance. *Nat. Rev. Microbiol*. **2009**, 7, 460–466.

45. Hoffman, L. R.; D'Argenio, D. A.; MacCoss, M. J.; Zhang, Z.; Jones, R. A.; Miller, S. I. Aminoglycoside antibiotics induce bacterial biofilm formation. *Nature* **2005**, 436, 1171–1175.

46. Segrè, D.; Deluna, A.; Church, G. M.; Kishony, R. Modular epistasis in yeast metabolism. *Nat. Genet*. **2005**, 37, 77–83.

47. Lamb, J.; Crawford, E. D.; Peck, D.; Modell, J. W.; Blat, I. C.; Wrobel, M. J.; Lerner, J.; Brunet, J.; Subramanian, A.; Ross, K. N.; et al. The Connectivity Map: using gene-expression signatures to connect small molecules, genes, and disease. *Science* **2006**, 313, 1929–1935.

48. Subramanian, A.; Tamayo, P.; Mootha, V. K.; Mukherjee, S.; Ebert, B. L.; Gillette, M. A.; Paulovich, A.; Pomeroy, S. L.; Golub, T. R.; Lander, E. S.; Mesirov, J. P. Gene set enrichment analysis: a knowledge-based approach for interpreting genome-wide expression profiles. *Proc. Natl. Acad. Sci. USA* **2005**, 102, 15545–15550.

49. Glaser, K. B.; Staver, M. J.; Waring, J. F.; Stender, J.; Ulrich, R. G.; Davidsen, S. K. Gene expression profiling of multiple histone deacetylase

(HDAC) inhibitors: defining a common gene set produced by HDAC inhibition in T24 and MDA carcinoma cell lines. *Mol. Cancer Ther*. **2003**, 2, 151–163.

50. Khalid, S. A.; Duddeck, H.; Gonzalez-Sierra, M. Isolation and characterization of an antimalarial agent of the neem tree *Azadirachta indica*. *J. Nat. Prod*. **1989**, 52, 922–926.
51. Fire, A.; Xu, S.; Montgomery, M. K.; Kostas, S. A.; Driver, S. E.; Mello, C. C. Potent and specific genetic interference by double-stranded RNA in *Caenorhabditis elegans*. *Nature* **1998**, 391, 806–811.
52. Stark, G. R.; Kerr, I. M.; Williams, B. R.; Silverman, R. H.; Schreiber, R. D. How cells respond to interferons. *Annu. Rev. Biochem*. **1998**, 67, 227–264.
53. Elbashir, S. M.; Harborth, J.; Lendeckel, W.; Yalcin, A.; Weber, K.; Tuschl, T. Duplexes of 21-nucleotide RNAs mediate RNA interference in cultured mammalian cells. *Nature* **2001**, 411, 494–498.
54. Mittal, V. Improving the efficiency of RNA interference in mammals. *Nat. Rev. Genet*. **2004**, 5, 355–365.
55. Lum, L.; Yao, S.; Mozer, B.; Rovescalli, A.; Von Kessler, D.; Nirenberg, M.; Beachy, P. A. Identification of hedgehog pathway components by RNAi in *Drosophila* cultured cells. *Science* **2003**, 299, 2039–2045.
56. Hsieh, A. C.; Bo, R.; Manola, J.; Vazquez, F.; Bare, O.; Khvorova, A.; Scaringe, S.; Sellers, W. R. A library of siRNA duplexes targeting the phosphoinositide 3-kinase pathway: determinants of gene silencing for use in cell-based screens. *Nucleic Acids Res*. **2004**, 32, 893–901.
57. Berns, K.; Hijmans, E. M.; Mullenders, J.; Brummelkamp, T. R.; Velds, A.; Heimerikx, M.; Kerkhoven, R. M.; Madiredjo, M.; Nijkamp, W.; Weigelt, B.; et al. A large-scale RNAi screen in human cells identifies new components of the p53 pathway. *Nature* **2004**, 428, 431–437.
58. Ong, S.; Schenone, M.; Margolin, A. A.; Li, X.; Do, K.; Doud, M. K.; Mani, D. R.; Kuai, L.; Wang, X.; Wood, J. L.; et al. Identifying the proteins to which small-molecule probes and drugs bind in cells. *Proc. Natl. Acad. Sci. USA* **2009**, 106, 4617–4622.
59. Wermuth, C. G. Selective optimization of side activities: the SOSA approach. *Drug Discov. Today* **2006**, 11, 160–164.
60. Scheiber, J.; Chen, B.; Milik, M.; Sukuru, S. C. K.; Bender, A.; Mikhailov, D.; Whitebread, S.; Hamon, J.; Azzaoui, K.; Urban, L.; et al. Gaining insight into off-target mediated effects of drug candidates with a comprehensive systems chemical biology analysis. *J. Chem. Inf. Model*. **2009**, 49, 308–317.
61. Yang, L.; Chen, J.; He, L. Harvesting candidate genes responsible for serious adverse drug reactions from a chemical–protein interactome. *PLoS Comput. Biol*. **2009**, 5, e1000441.
62. Keiser, M. J.; Setola, V.; Irwin, J. J.; Laggner, C.; Abbas, A. I.; Hufeisen, S. J.; Jensen, N. H.; Kuijer, M. B.; Matos, R. C.; Tran, T. B.; et al. Predicting new molecular targets for known drugs. *Nature* **2009**, 462, 175–181.

63. Campillos, M.; Kuhn, M.; Gavin, A.; Jensen, L. J.; Bork, P. Drug target identification using side-effect similarity. *Science* **2008**, 321, 263–266.

64. Karni, S.; Soreq, H.; Sharan, R. A network-based method for predicting disease-causing genes. *J. Comput. Biol*. **2009**, 16, 181–189.

65. Qu, X. A.; Gudivada, R. C.; Jegga, A. G.; Neumann, E. K.; Aronow, B. J. Inferring novel disease indications for known drugs by semantically linking drug action and disease mechanism relationships. *BMC Bioinf*. **2009**, 10(Suppl. 5), S4.

66. Lehár, J.; Krueger, A.; Zimmermann, G.; Borisy, A. High-order combination effects and biological robustness. *Mol. Syst. Biol*. **2008**, 4, 215.

67. Lehár, J.; Zimmermann, G. R.; Krueger, A. S.; Molnar, R. A.; Ledell, J. T.; Heilbut, A. M.; Short, G. F.; Giusti, L. C.; Nolan, G. P.; Magid, O. A.; et al. Chemical combination effects predict connectivity in biological systems. *Mol. Syst. Biol*. **2007**, 3, 80.

INDEX

Computational Approaches in Cheminformatics and Bioinformatics, First Edition.
Edited by Rajarshi Guha and Andreas Bender.